When
LEAST
Is **Best**

When

LEAST

PRINCETON UNIVERSITY PRESS
Princeton and Oxford

PAUL J. NAHIN

How Mathematicians Discovered
Many Clever Ways to Make Things
as Small (or as Large) as Possible

Is Best

With a new preface by the author

Copyright © 2004 by Princeton University Press
Published by Princeton University Press,
41 William Street, Princeton, New Jersey 08540

In the United Kingdom: Princeton University Press,
3 Market Place, Woodstock,
Oxfordshire OX20 1SY

Fifth printing, and first paperback printing, with a new
preface by the author, 2007
Paperback ISBN-13: 978-0-691-13052-1

The Library of Congress has cataloged the cloth edition
of this book as follows

Nahin, Paul J.
    When least is best : how mathematicians discovered
many clever ways to make things as small (or as large) as
possible / Paul J. Nahin.
        p. cm.
    Includes bibliographical references and index.
    1. Maxima and minima.   2. Mathematics—History.
I. Title.
QA306.N34 2004
511'.66—dc22   2003055537

British Library Cataloging-in-Publication Data is available

This book has been composed in Stone

Printed on acid-free paper. ∞

press.princeton.edu

Printed in the United States of America

10 9 8 7 6

For **PATRICIA ANN**,

who contradicts the title because she

has always been

"the most" and is

still the **best**

When a quantity is the greatest or the least that it can be, at that moment it neither flows backwards nor forwards; for if it flows forwards or increases it was less, and will presently be greater than it is; and on the contrary if it flows backwards or decreases, then it was greater, and will presently be less than it is.

— **Isaac Newton** on maximums and minimums, in *Methodus fluxionum et serierum infinitarum*, 1671

There are hardly any speculations in geometry more useful or more entertaining than those which relate to maxima and minima.

— the great English mathematician **Colin Maclaurin**, in *A Treatise of Fluxions*, 1742

The great body of physical science, a great deal of the essential fact of financial science, and endless social and political problems are only accessible and only thinkable to those who have had a sound training in mathematical analysis, and the time may not be very remote when it will be understood that for complete initiation as an efficient citizen of one of the great complex world-wide States that are now developing, it is as necessary to be able to compute, to think in averages and maxima and minima, as it is now to be able to read and write.

— **H. G. Wells**, from *Mankind in the Making*, 1903

Contents

$\mathbf{P}$reface to the Paperback Edition

All the greatest mathematicians have long
since recognized that the method presented
in this book is not only extremely useful in
analysis, but that it also contributes greatly to
the solution of physical problems. For since
the fabric of the universe is most perfect, and
is the work of a most wise Creator, nothing
whatsoever takes place in the universe in
which some relation of maximum and
minimum does not appear.

—the great Swiss-born mathematician **Leonhard
Euler**, in *Methodus inveniendi lineas curvas
maximi minimive proprietate gaudentes, sive
solutio problematis isoperimetrici lattissimo
sensu accepti,* 1744*

In a letter dated October 11, 1709, the well-known English
scientist Roger Cotes wrote to his even better-known friend Isaac
Newton. Cotes, who was in charge of preparing the second edition
of Newton's monumental *Principia* for publication, had a gloomy
message to deliver, stating "It is impossible to print the book without
some faults." Events proved him to be correct. After the appearance
of the second edition, Newton sent Cotes a list of new corrections,
which prompted Cotes to reply, in a letter dated December 22, 1713,
"I observe you have put down 20 Errata. . . . I believe you will not
be surprised if I tell you I can send you 20 more." Cotes then went
on to reveal that while he was preparing the second edition he had

---

* In English, A Method for Finding Curved Lines Enjoying Properties of Maximum
or Minimum, or Solution of Isoperimetric Problems in the Broadest Accepted Sense.
The quotation is taken from the opening paragraph of the book's appendix, which
is titled "De curvis elasticis." You can find a complete, annotated English translation of the appendix in W. A. Oldfather et al., "Leonhard Euler's Elastic Curves,"
Isis 20, no. 1 (November 1933): 72–160.

"made some hundreds [of additional corrections to the first edition] with which I never acquainted you."

Well, this book isn't the *Principia* (and I'm no Newton), but it does share one characteristic with that genius's masterpiece—the first editions of both books had some errors! Not quite so many in this book as Cotes mentioned, I think, but a few. The appearance of the paperback edition of *When Least Is Best* has given me the opportunity to make those missteps go away, and I gratefully thank Vickie Kearn, my longtime editor at Princeton University Press, for that opportunity.

Besides typographical errors, there were two errors of citation omission that I would like to now correct. First, the discussion on pages 28 through 33 was motivated when I read the paper by Nathaniel Silver, "A Refraction Problem in Several Variables," *American Mathematical Monthly*, June-July 1987: 545–47. And second, the perfect basketball shot discussion on pages 158 through 165, although presented as a natural spin-off of Halley's gunnery problem (for which I cited a 1997 paper by C. W. Groetsch) was actually discussed sixteen years before the appearance of Professor Groetsch's more general, historical paper, in an analysis by G. J. Porter, "New Angles on an Old Game," *American Mathematical Monthly*, April 1981: 285–86.

In the discussion on pages 56 through 60, on Jacob Steiner's flawed geometric proof of the isoperimetric theorem, I make reference to Besicovitch's solution to Kakeya's problem. I make only some brief, general comments on what Besicovitch proved, but you can find much more in two papers: "On a Theorem of Besicovitch" by Hans Rademacher, and "On the Besicovitch-Perron Solution to the Kakeya Problem," both of which are in *Studies in Mathematical Analysis and Related Topics: Essays in Honor of George Pólya*, edited by Gábor Szegö et al. (Stanford University Press, 1962). In the second paper, the "Perron" is the German mathematician Oskar Perron (1880–1975), who in 1913 formulated an amusing "proof" to illustrate the flaw in Steiner's isoperimetric proof. As I discuss in the text, Steiner made his error right at the start, with his assumption that there actually is a closed curve of given length that encloses the maximum area. Assuming that an extrema question actually has an answer can lead one astray, however, as it does in dramatic fashion in *Perron's paradox*, which is a "proof" that 1 is the largest integer!

Here's how it goes: start by assuming that there is in fact an $N > 1$ that is the largest integer. Then, N^2 is an integer and, of course, $N^2 > N$, which is in conflict with the assumption that N is the largest integer. Therefore our starting assumption that $N > 1$ must be wrong and so it must be true that $N = 1$.

Now I'd be willing to bet that all readers of this book know that the proper concluding statement should actually be that the starting assumption that there actually *is* a largest integer is wrong, i.e., that the assumption that we can actually determine the largest integer is wrong. This is because we know how integers "work"—there is no largest one because there is always a bigger one, no matter how big the one we think of is. Just add one! And that's the whole point to Perron's paradox, of course; in those problems where we really don't know *a priori* how things "work," the *assumption* of the existence of a solution might well lead us into disaster.

On page 259 there is a challenge problem for you to consider, based on the isoperimetric theorem, the proof of which has just been completed on the preceding pages. As I explain there, I don't know how to solve that challenge problem, and in the first edition I asked readers to send me a solution if they had success. You can read the details of the challenge problem on page 259, but for now let me just say that the problem is that of finding a derivation of the (claimed) inequality

$$\int_0^{2\pi} \sqrt{a^2 \sin^2(t) + b^2 \cos^2(t)}\,dt \geq \sqrt{4\pi\{\pi ab + (a-b)^2\}},$$

where a and b are non-negative (but otherwise arbitrary) constants. This inequality is arrived at in this book by purely geometric arguments—the challenge (for you) is to find an *analytical* derivation.

I received just three letters. The first, from a reader in Pennsylvania, claimed to have a proof. But it was simply a demonstration that if an ellipse and a circle have the same perimeter, then the area of the ellipse is no greater than that of the circle. It was a clever bit of analysis, but of course, while true, it is just a special case of the isoperimetric theorem, which is proven in the book just before I state the challenge problem. The second letter (whose author did properly understand what was to be shown) was from the other end of the spectrum; it was from a physicist in Scotland who asserted that the claimed inequality is "false and [so] there is no possible derivation for it." He believed that my reasoning in arriving at the inequality contained "a deep flaw," and that I had been led astray by "one of those rare situations where the crazy world of topology intrudes into the 'real' world in a 'visible' way." Since all I do in the book is cut an ellipse into four good-sized parts and then rearrange them, I found his assertion to be just a bit hard to accept. Now, one might counter my reaction by observing that a simple half-twist to a long strip of paper, followed by joining the two distant ends of the strip, turns a two-sided object (the original strip) into a loop with a *single* side (the famous Möbius band) and that certainly *is* a bizarre topological intrusion. Perhaps my shuffling of the ellipse pieces had done something equally weird. There was a lot of hand-written mathematical analysis included in the physicist's letter to back up his words, and although it was clearly the work of an intelligent author, I was reluctant to devote what I was sure would be a time-consuming effort to wade through it.

But, what if he was right? It wouldn't be the first time I had made a mistake!

I decided to follow his suggestion that "perhaps the best check [of the inequality] would be to look at numerical values of the integral and compare with calculated values of $\sqrt{4\pi\{\pi ab + (a-b)^2\}}$." He admitted that he had not done that. I, on the other hand, keep a hot-to-trot MATLAB application idling on my computer's desktop 24/7. After all, one never knows when a number-crunching emergency might occur—and if there ever was such an emergency, this was it! It was duck soup to write the brief code to do what the physicist suggested, and here's how I proceeded.

The claimed inequality can be slightly altered as follows:

$$b\int_0^{2\pi} \sqrt{\left(\frac{a}{b}\right)^2 \sin^2(t) + \cos^2(t)}\, dt \geq b\sqrt{4\pi\left\{\pi\frac{a}{b} + \left(\frac{a}{b} - 1\right)^2\right\}}.$$

Or, writing $x = a/b$, where $0 \leq x < \infty$, the challenge problem is equivalent to analytically deriving the following inequality (valid for all non-negative x):

$$\int_0^{2\pi} \sqrt{x^2 \sin^2(t) + \cos^2(t)} \, dt - \sqrt{4\pi \{\pi x + (x-1)^2\}} \geq 0.$$

Before studying the "truth" of this inequality by computer, there are two special cases we can use to partially check the MATLAB coding.

For $x = 0$ the claim becomes

$$\int_0^{2\pi} \sqrt{\cos^2(t)} \, dt - \sqrt{4\pi} \geq 0,$$

that is,

$$\int_0^{2\pi} |\cos(t)| \, dt - 2\sqrt{\pi} \geq 0.$$

(Notice, *carefully*, that $\int_0^{2\pi} \sqrt{\cos^2(t)} \, dt \neq \int_0^{2\pi} \cos(t) \, dt$.) Now, since $\cos(t) \geq 0$ for $0 \leq t \leq \pi/2$, our claim becomes

$$4 \int_0^{\pi/2} \cos(t) \, dt - 2\sqrt{\pi} \geq 0$$

or,

$$4\{ \sin(t)|_0^{\pi/2} - 2\sqrt{\pi} \geq 0$$

or,

$$4 - 2\sqrt{\pi} = 0.4551 \geq 0,$$

which is, of course, true (even easier is to just recall that $\pi < 4$). For $x = 1$ the claim becomes

$$\int_0^{2\pi} \sqrt{\sin^2(t) + \cos^2(t)} \, dt - \sqrt{4\pi^2} \geq 0$$

or,

$$\int_{0}^{2\pi} dt - 2\pi \geq 0$$

or,

$$2\pi - 2\pi = 0 \geq 0,$$

which is, of course, true. The results from our MATLAB code should be consistent with these two particular calculations.

Figure P shows the left-hand side of the boxed inequality, as a function of x, over the interval $0 \leq x \leq 5$. The plot agrees in particular with our two special cases above, and in general with the inequality, because the curve never dips below the x-axis. Because of this plot, I conclude it is safe to say that the challenge problem still

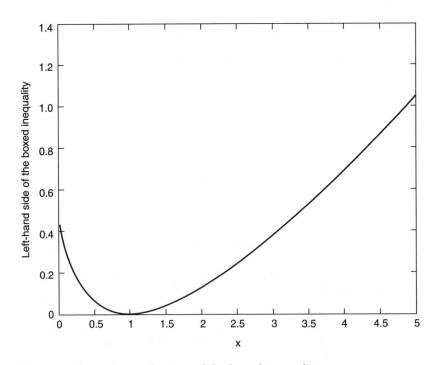

FIGURE P. Computer verification of the boxed inequality.

stands. And, I should tell you, two weeks after his first letter arrived, I got a second letter from the Scottish physicist telling me, in so many words, "oops": he had found a fatal error in his original analysis and had graciously written to say "your result . . . stands unchallenged!"

Now, as long as I've been discussing a challenge problem, let me give you a new one to consider as you read the paperback edition of the book.

New Challenge Problem for the Paperback Edition

Let a, b, and c be the lengths of the sides of any triangle. Show that

$$abc \geq (a + b - c)(b + c - a)(a + c - b).$$

Hint: You may find it helpful to recall the following result from high school plane geometry (it was known to Euclid and appears as a proposition in his *Elements*): the bisector lines of any triangle's three interior vertex angles cross at a *common point* called the *incenter* (so named because that point is guaranteed to be *in*side the triangle). For now I'll let you see if you can prove this for yourself, too; but if you can't, its (elementary) proof is included in appendix I, along with a derivation of the challenge inequality.

For the specific cases of the 45°-45°-90° and the 30°-60°-90° triangles, you can verify by direct calculation that the challenge inequality reduces to the claims that $9 > 8$ and $4 > 3$, respectively, which are both (of course!) true.

I'll conclude on the somber note that the mathematicians Leonid Khachiyan and George Dantzig, who appear in this book's chapter 7 discussion of linear programming, have both died since the publication of the hardcover. Remarkably, their obituary notices appeared on the same day, next to each other, in the *New York Times* (May 23, 2005, p. A17). Dantzig's long, productive life ended at age ninety, while Khachiyan's was cut tragically short by a heart attack at the young age of only fifty-two.

Lee, New Hampshire
January 24, 2007

Preface

This is a history of mathematics book, but it is *not* simply a collection of biographical, prose essays on the lives of various mathematicians. There is a place for that sort of book (e.g., see E. T. Bell's classic *Men of Mathematics*), but this isn't one of them. What it *is* is the technical story of what many brilliant mathematicians have done in the subject of extrema over the last two dozen centuries. To be blunt, there is *a lot of mathematics* in this book. Stephen Hawking's famous line about how every equation cuts a book's readership in half doesn't apply here—that's for coffee table books, ones more for displaying than for reading. *This* book is for readers with calluses on their fingers because they read with pencil and paper in hand!

While I hope much of what you read here will be new and exciting to you, I do expect you to bring some intellectual background to the table. In general, what a science or engineering major learns in the first year of undergraduate calculus and physics is pretty much enough (I'll be specific in the next paragraph). Actually, as far as the physics goes, all you really need to remember is that force is a vector, and what potential and kinetic energy are. For the math, however, there is a list of things I am assuming that is just a bit longer. First of all, do you find the following question easy? If we assume x is real, then what is

$$\lim_{n \to \infty} \frac{\sin(x)}{n} = ?$$

The answer is, of course, zero, since no matter what x may be the value of $\sin(x)$ is always in the interval -1 to $+1$, and so as $n \to \infty$ the expression goes to zero. Now, when I ask even my second-year engineering students this, I almost always get the correct answer of zero, and so they are astounded when I tell them the answer is really 6! Then I write on the blackboard, without saying a word,

$$\lim_{n \to \infty} \frac{\sin(x)}{n} = \lim_{n \to \infty} \frac{\sin(x)}{\not{n}} = \text{si}(x) = 6.$$

If they laugh at this astonishing "calculation" then I take that as a *good* sign—only a student who has reached a certain level of skill and knowledge would find the above to be *so wrong* as to be funny.

More seriously, I am assuming that you have a good background in high school algebra, trigonometry, and geometry, as well as in the elementary integration techniques of freshman calculus. For example, I am assuming that it will be unnecessary for me to explain the quadratic formula, or what it means to quote a trig identity, or what solid angles, hyperbolic functions, and factorials are (and that $0! = 1$, not 0), or what it means to say a real number is irrational, or what a vector is, or what the Pythagorean theorem is, or what it means to prove something by induction, or how to differentiate and integrate "simple" functions. On this last point, I expect you to know that not only is

$$\int_0^1 x \, dx = \frac{1}{2}$$

but also, without actually doing the integrals, that we can write

$$\int_1^8 \sin^{17}(x + \sqrt{x}) \, dx = \int_1^8 \sin^{17}(y + \sqrt{y}) \, dy.$$

I will assume, finally, that the physical interpretation of an integral as the area under a curve is a familiar one.

It might appear just a bit odd for me to assume you already know what a derivative is, since the evolution of that concept is a major part of chapter 4 in this book. But not to make that assumption is awkward; there will be places in the first three chapters where, to make a point, I'll want to compare a noncalculus calculation with one using differentiation. If you know what an integral is, then assuming you know what a derivative is seems (to me) to be logical.

As a more specific (and more interesting) example of what I have in mind, I am assuming that the following little analysis, while perhaps astonishing in its conclusion, also is understandable. One of

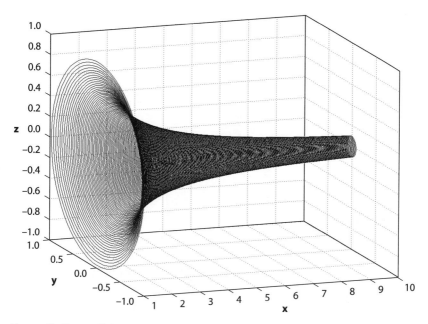

FIGURE I. Torricelli's paradoxical funnel.

the mathematical gems of seventeenth-century mathematics was the discovery of a surface of revolution that, even though infinite in extent, nevertheless bounds a finite volume. Prior to this discovery it was commonly accepted that a surface extended to infinity would *necessarily* have to be of infinite size, i.e., enclose infinite volume. In contradiction to that common belief, in 1641 the Italian mathematical physicist Evangelista Torricelli (1608–47) discovered that if the first quadrant branch of the hyperbola $xy = 1$ (with $x \geq a$) is rotated about the x-axis, as shown in Figure I with $a = 1$, then the resulting surface (resembling an infinitely long horn, sometimes called *Gabriel's horn*, after the Biblical story of the archangel who blew it before making an announcement) bounds finite volume.

The demonstration of this result is technically quite simple (which is why I am assuming you will be able to follow the details). We first imagine that the volume is sliced up into arbitrarily many thin cylindrical disks, each with thickness Δx. The radius of a disk with its center at x is $y = 1/x$. Thus, the volume of that disk is, from solid geometry, approximately given by

$$\Delta V \approx \pi y^2 \Delta x = \frac{\pi}{x^2} \Delta x.$$

As $\Delta x \to 0$, we can replace ΔV and Δx with the differentials dV and dx, respectively; our approximation becomes exact, and so we have

$$dV = \pi \frac{dx}{x^2}.$$

To find the total volume V we simply integrate from $x = a$ to infinity, and so

$$V = \int dV = \pi \int_a^\infty \frac{dx}{x^2} = \pi \left\{ -\frac{1}{x} \Big|_a^\infty \right\} = \frac{\pi}{a}.$$

Torricelli's result was thought paradoxical in the years following its announcement—see Paolo Mancosu and Ezio Vailati, "Torricelli's Infinitely Long Solid and Its Philosophical Reception in the Seventeenth Century" (*Isis*, March 1991, pp. 50–70); in 1672 the English philosopher Thomas Hobbes declared that one would have to be crazy to believe Torricelli. (If Hobbes' philosophical powers had been equal to his mathematical skills no one would remember him today.) Even today Torricelli's calculation can provoke a great deal of discussion in freshman calculus classrooms. Consider, for example, the fact that the area A of the Torricelli surface of revolution is infinite, a result easily confirmed by calculating the value of the area integral:

$$A = \int_a^\infty y \sqrt{1 + \left(\frac{dy}{dx} \right)^2} \, dx.$$

(You can find this general formula for the surface area of the rotated curve $y = y(x)$ discussed in section 6.9). Thus, since $y = 1/x$, we have

$$\frac{dy}{dx} = -\frac{1}{x^2},$$

and so

$$A = \int_a^\infty \frac{1}{x} \sqrt{1 + \frac{1}{x^4}} \, dx = \int_a^\infty \frac{\sqrt{x^4 + 1}}{x^3} \, dx > \int_a^\infty \frac{\sqrt{x^4}}{x^3} \, dx = \int_a^\infty \frac{dx}{x}.$$

The inequality follows as I have replaced the numerator of the integral with an expression that is, for every x in the interval of integration, smaller than the exact numerator, i.e., $x^4 < x^4 + 1$. But the last integral diverges logarithmically and so A is infinite. This would appear to mean that we could *not* paint the interior of Torricelli's surface, as we would require an infinite amount of paint to cover an infinite area. And yet we *could* paint the interior by simply filling the *finite* volume with paint. We seem to have a paradox.

Well, even if you are now mumbling to yourself over whether or not we can paint Torricelli's surface, if the mathematics itself is understandable to you then you know all of the mathematics you need to know to start reading this book. (Can you see how to resolve this paradox? The answer is at the end of this preface, but don't look until you've thought about it for a while. Hint: there is a difference between *real* paint and *mathematical* paint!)

Finally, to conclude this little essay, to read this book for the maximum gain you should have something that professors like to call "mathematical maturity." This is an attribute intentionally left fuzzy, meant only to describe a "mind ready to receive" new material (or perhaps old material in a new way). *My* quick test for mathematical maturity is to see how a student reacts to the following little gem of reasoning.

Recall that all real numbers can be separated into one of two sets—the *rationals* (expressible as m/n, the *ratio* of two integers), and the *irrationals* (all the remaining reals which are, of course, *not* rational). Every real number is one or the other. An example of each is $0.3333\cdots = \frac{1}{3}$, and $\sqrt{2}$, respectively. In some sense, then, the irrationals have a "more complicated" structure than the rationals. Now, here's the problem: if we raise an irrational number to an irrational power, can the result be rational? Most students come down on the side of *no*, arguing that combining two irrationals through the power operation is "messy" and seems incapable of producing something simple like m/n. But then I show them it *is* possible, and watch how they respond.

Start by considering $(\sqrt{2})^{\sqrt{2}}$, an irrational number raised to an irrational power. It is itself a real number and so it is either rational or it is irrational (we do *not* have to actually calculate it). If it is rational, we are done. If it is irrational, then raise *it* to an irrational power, e.g., consider $((\sqrt{2})^{\sqrt{2}})^{\sqrt{2}}$. But this is $(\sqrt{2})^2 = 2$, which is rational and so, again, we are done. Notice that we still don't know from this argument if $(\sqrt{2})^{\sqrt{2}}$ is rational or not—and it doesn't matter! (It has been known since 1930 that $(\sqrt{2})^{\sqrt{2}}$ is not only irrational, but transcendental.)

To judge mathematical maturity, I look for two things. First, of course, is simply that there is a technical understanding of the argument. But I also want to see *excitement*, a "Wow, what a *neat* proof!" reaction. For me, that's the signature of a mind "ready to receive." I hope that was your reaction to the above, and that this book will give you lots more (indeed, a *maximum*) of "Wow, that's neat!" moments.

Painting Torricelli's Funnel

The reason for the "paradox" is that you are simultaneously holding two contradictory ideas about the nature of paint. Real paint has a molecular structure, i.e., there is a smallest (nonzero) volume of real paint, while mathematical paint is infinitely divisible. Consistently using *either one* of these two conceptions of paint removes the paradox. Here's how.

For mathematical paint: We can indeed paint the funnel's inner surface by simply filling the funnel with a finite volume of paint. It does *not* follow, however, that it takes an infinite volume of mathematical paint to cover an infinite surface area, since the thickness of mathematical paint is zero. That is, infinite area times zero thickness is an *indeterminate* volume. The paradox has vanished.

For real paint: It would, indeed, require an infinite volume of real paint to paint the *outer* surface of the funnel because real paint has a nonzero thickness. But it is impossible to paint the entire inner surface (equal in area, of course, to the outer surface area) because the paint won't fit! At some point along the ever narrowing funnel, the opening will be smaller than a single molecule of real paint. This means we simply cannot compare the two different ways of painting the funnel since filling the funnel with real paint cannot even be done.

We have a "paradox" only if we imagine filling the funnel with mathematical paint but painting the outer surface with real paint. Getting a paradox by changing the rules in "mid-game" is no surprise at all.

When
LEAST
Is **Best**

1.

Minimums, Maximums, Derivatives, and Computers

1.1 Introduction

This book has been written from the practical point of view of the engineer, and so you'll see few rigorous proofs on any of the pages that follow. As important as such proofs are in modern mathematics, I make no claims for rigor in this book (plausibility and/or direct computation are the themes here), and if absolute rigor is what you are after, well, you have the wrong book. Sorry!

Why, you may ask, are *engineers* interested in minimums? That question could be given a very long answer, but instead I'll limit myself to just two illustrations (one serious and one not, perhaps, quite as serious). Consider first the problem of how to construct a gadget that has a fairly short operational lifetime and which, during that lifetime, must perform flawlessly. Short lifetime and low failure probability are, as is often the case in engineering problems, potentially conflicting specifications: the first suggests using low-cost material(s) since the gadget doesn't last very long, but using cheap construction may result in an unacceptable failure rate. (An example from everyday life is the ordinary plastic trash bag—how thick should it be? The bag is soon thrown away, but we definitely will be unhappy if it fails too soon!) The trash bag engineer needs to calculate the minimum thickness that still gives acceptable performance.

For my second example, let me take you back to May 1961, to the morning the astronaut Alan Shepard lay on his back atop the rocket that would make him America's first man in space. He was very brave to be there, as previous unmanned launches of the same type of rocket had shown a disturbing tendency to explode into stupendous fireballs. When asked what he had been thinking just before blastoff, he replied "I was thinking that the whole damn thing had been built by the lowest bidder."

This book is a math history book, and the history of minimums starts centuries before the time of Christ. So, soon, I will be starting at the beginning of our story, thousands of years in the past. But before we climb into our time machine and travel back to those ancient days, there are a few modern technical issues I want to address first.

First, to write a book on minimums might seem to be a bit narrow; why not include maximums, too? Why not write a history of *extremas*, instead? Well, of course minimums and maximums are indeed certainly intimately connected, since a maximum of $y(x)$ is a minimum of $-y(x)$. To be honest, the reason for the book's title is simply that I couldn't think of one I could use with extrema as catchy as is "When Least Is Best." I did briefly toy with "When Extrema Are xxx" with the xxx replaced with *exotic, exciting,* and even (for a while, in a temporary fit of marketing madness that I hoped would attract Oprah's attention), *erotic.* Or even "Minimums Are from Venus, Maximums Are from Mars." But all of those (certainly the last one) are dumb, and so it stayed "When Least Is Best." There will be times, however, when I will discuss maximums, too. And now and then we'll use a computer as well.

For example, consider the problem of finding the maximum value of the rather benign-looking function

$$y(x) = 3\cos(4\pi x - 1.3) + 5\cos(2\pi x + 0.5).$$

Some students answer too quickly and declare the maximum value is 8, believing that for some value of x the individual maximums of the two cosine terms will add. That is not the case, however, since it is equivalent to saying that there is some $x = \hat{x}$ such that

$$4\pi\hat{x} - 1.3 = 2\pi n$$

$$2\pi\hat{x} + 0.5 = 2\pi k,$$

where n and k are integers. That is, those students are assuming there is an $\hat{x}$ such that

$$\hat{x} = \frac{2\pi n + 1.3}{4\pi} = \frac{2\pi k - 0.5}{2\pi}, \qquad n \text{ and } k \text{ integers.}$$

Thus,

$$2n\pi + 1.3 = 4\pi k - 1,$$

or

$$2.3 = 4\pi k - 2\pi n = 2\pi(2k - n),$$

or

$$\pi = \frac{2.3}{2(2k - n)} = \frac{23}{20(2k - n)}.$$

But if this is actually so, then as n and k are integers we would have π as the ratio of integers, i.e., π would be a rational number. Since 1761, however, π has been known to be irrational and so there are no integers n and k. And that means there is no $\hat{x}$ such that $y(\hat{x}) = 8$, and so $y_{max}(x) < 8$.

Well, then, what is $y_{max}(x)$? Is it perhaps close to 8? You might try setting the derivative of $y(x)$ to zero to find $\hat{x}$, but that quickly leads to a mess. (Try it.) The best approach, I think, is to just numerically study $y(x)$ and watch what it does. The result is that $y_{max}(x) = 5.7811$, significantly less than 8. My point in showing you this is twofold. First, a computer is often quite useful in minimum studies (and we will use computers a lot in this book). Second, taking the derivative of something and setting it equal to zero is not always what you have to do when finding the extrema of a function.

An amusing (and perhaps, for people who like to camp, even useful) example of this is provided by the following little puzzle. Imagine that you have been driving for a long time along a straight road that borders an immense, densely wooded area. It looks enticing, and so you park your car on the side of the road and hike into the woods for a mile along a straight line perpendicular to the road. The woods are very dense (you instantly lose sight of the road when you are just one step into the woods), and after a mile you are exhausted.

You call it a day and camp overnight. When you get up the next morning, however, you've completely lost your bearings and don't know which direction to go to get back to your car. You could, if you panic, wander around in the woods indefinitely! But there *is* a way to travel that absolutely guarantees that you will arrive back at your car's *precise location* after walking a certain maximum distance (it might take even less). How do you walk out of the woods, and what is the maximum distance you would have to walk? The answer requires only simple geometry—if you are stumped the answer is at the end of this chapter.

1.2 When Derivatives Don't Work

Here's another example of a minimization problem for which calculus is not only *not* required, but in fact seems not to be able to solve. Suppose we have the real line before us (labeled as the x-axis), stretching from $-\infty$ to $+\infty$. On this line there are marked n points, labeled in increasing value as $x_1 < x_2 < \cdots < x_n$. Let's assume all the x_i are finite (in particular x_1 and x_n), and so the interval of the x-axis that contains all n points is finite in length. Now, somewhere (any-where) on the finite x-axis we mark one more point (let's call it x). We wish to pick x so that the sum of the distances between x and all of the original points is minimized. That is, we wish to pick x so that

$$S = |x - x_1| + |x - x_2| + \cdots + |x - x_n|$$

is minimized. I've used absolute-value signs on each term to insure each *distance* is non-negative, independent of where x is, either to the left or to the right of a given x_i. Those absolute-value signs may seem to badly complicate matters, but that's not so. Here's why.

First, focus your attention on the two points that mark the ends of the interval, x_1 and x_n. The sum of the distances between x and x_1, and between x and x_n, is

$$|x - x_1| + |x - x_n|$$

and this is *at least* $|x_1 - x_n|$. If $x > x_n$, or if $x < x_1$ (i.e., if x is outside the interval), then strict *in*equality holds, but if x is *anywhere* inside the interval (i.e., $x_1 \leq x \leq x_n$) then equality holds. Thus, the minimum value of $|x-x_1|+|x-x_n|$ is achieved by placing x anywhere between x_1 and x_n.

Next, shift your attention to the two points x_2 and x_{n-1}. We can repeat the above argument, without modification, to conclude that the minimum value of $|x - x_2| + |x - x_{n-1}|$ is achieved when x is *anywhere* between x_2 and x_{n-1}. Note that this automatically satisfies the condition for minimizing the value of $|x - x_1| + |x - x_n|$, i.e., placing x anywhere between x_2 and x_{n-1} minimizes $|x - x_1| + |x - x_2| + |x - x_{n-1}| + |x - x_n|$. You can now see that we can repeat this line of reasoning, over and over, to conclude

$$|x - x_3| + |x - x_{n-2}| \quad \text{is minimized by placing } x \text{ } anywhere$$
$$\text{between } x_3 \text{ and } x_{n-2},$$

$$|x - x_4| + |x - x_{n-3}| \quad \text{is minimized by placing } x \text{ } anywhere$$
$$\text{between } x_4 \text{ and } x_{n-3},$$

$$\vdots$$

and finally, if we suppose that n is an *even* number of points, then

$$|x - x_{\frac{n}{2}}| + |x - x_{\frac{n}{2}+1}| \quad \text{is minimized by placing } x \text{ } anywhere$$
$$\text{between } x_{\frac{n}{2}} \text{ and } x_{\frac{n}{2}+1}.$$

So, we simultaneously satisfy all of these individual minimizations by placing x anywhere between $x_{n/2}$ and $x_{(n/2)+1}$ (*if* n is even), and this of course minimizes S.

But what if n is odd? Then the same reasoning as for even n still works, *until* the final step; then there is no second point to pair with $x_{(n+1)/2}$. Thus, simply let $x = x_{(n+1)/2}$ and so $|x - x_{(n+1)/2}| = 0$, which is certainly the minimum value for a *distance*. Thus, we have the somewhat unexpected, noncalculus solution that, for n even, S is minimized by placing x *anywhere* in an *interval*, but for n odd there is just *one, unique* value for x (the *middle* x_i) that minimizes S.

1.3   Using Algebra to Find Minimums

As another elementary but certainly not a trivial example of the claim that derivatives are not always what you want to calculate, consider the fact that ancient mathematicians knew that of all rectangles with a given perimeter it is the square that has the largest area. (This is a special result from a general class of maximum/minimum questions of great historical interest and practical value called *isoperimetric problems*, and I'll have more to say about them in the next chapter.) Ask most modern students to show this and you will almost surely get back something like the following. Define P to be the given perimeter of a rectangle, with x denoting one of the two side lengths. The other side length is then $(P - 2x)/2$, and so the area of the rectangle is

$$A(x) = x \left(\frac{P - 2x}{2} \right) = \frac{1}{2} Px - x^2.$$

$A(x)$ is maximized by setting $dA/dx = \frac{1}{2}P - 2x$ equal to zero, and so $x = \frac{1}{4}P$, which completes the proof. Using only algebra, however, an ancient mathematician could have argued that

$$A = x \left(\frac{P - 2x}{2} \right) = \frac{1}{2} Px - x^2 = \frac{P^2}{16} - \frac{P^2}{16} + \frac{1}{2} Px - x^2$$

$$= \frac{P^2}{16} - \left(x^2 - \frac{1}{2} Px + \frac{P^2}{16} \right) = \frac{P^2}{16} - \left(x - \frac{P}{4} \right)^2 \leq \frac{P^2}{16}$$

since $(x - (P/4))^2 \geq 0$ for all x. That is, A is never larger than the constant $P^2/16$ and is equal to $P^2/16$ if and only if (a useful phrase I will henceforth write as simply iff) $x = P/4$, which completes the ancient, noncalculus proof.

As a final comment on this result, which again illustrates the intimate connection between minimum and maximum problems, we can restate matters as follows: of all rectangles with a given area, the square has the smallest perimeter. This is the so-called *dual* of our original problem and, indeed, all isoperimetric problems come in such pairs. I'll prove this particular dual in section 1.5. Another useful isoperimetric result that seems much like the one just established—one also known to the precalculus, ancient mathematicians—is not so easy to prove: of all the triangles with the same area, the equilateral has the smallest perimeter. See if you can show this (or its dual) before I do it later in this chapter.

We can use the previous result—of all rectangles with a fixed perimeter, the square has the maximum area—to solve *without* calculus a somewhat more complicated appearing problem found in all calculus textbooks. Suppose we wish to enclose a rectangular plot of land with a fixed length of fencing, with the *side of a barn* forming one side of the enclosure. How should the fencing now be used? We could, of course, use calculus as follows: let x be the length of each of the two sides perpendicular to the barn wall, and $\ell - 2x$ be the length of the side parallel to the barn wall (ℓ is the fixed, total length of the fencing). Then the enclosed area is

$$A = x(\ell - 2x) = x\ell - 2x^2$$

and so

$$\frac{dA}{dx} = \ell - 4x,$$

which, when set equal to zero, gives $x = \frac{1}{4}\ell$. Thus, $\ell - 2x = \frac{1}{2}\ell$, which says the enclosed area is maximized when it is twice as long as it is wide. But this solution is far more sophisticated than required. Simply imagine that we enclose another rectangular area *on the other side* of the barn wall. We already know that, together, the two rectangular plots should form a square, and so each of the two rectangular plots are *half* of the square, i.e., twice as long in one dimension as in the other.

Our ancient mathematician's trick of completing the square is a very old one, and some historians claim that it can be found implicit in Euclid's *Elements* (Book 6, Proposition 27), circa 300 B.C. There, the problem discussed is equivalent to that of dividing a constant into two parts so that their product is maximum. So, if the constant is C, then the two parts are x and $C - x$, with the product

$$M = x(C - x) = Cx - x^2 = -(x^2 - Cx)$$

$$= -\left(x^2 - Cx + \frac{C^2}{4} - \frac{C^2}{4}\right) = -\left(x - \frac{C}{2}\right)^2 + \frac{C^2}{4}.$$

Thus, as $(x - (C/2))^2 \geq 0$ for all x, then M is never larger than $C^2/4$ and is equal to $C^2/4$ iff $x = C/2$.

Stated this way, Euclid's problem surely seems rather abstract, but in 1673 the Dutch mathematical physicist Christiaan Huygens gave a nice physical setting to the calculation. Suppose we have a line and

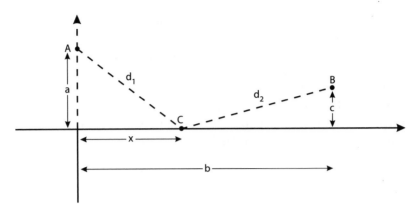

FIGURE 1.1. Huygen's problem.

two points (A and B) *not* on the line. Where should the point C be located *on* the line so that the sum of the squares of the distances from C to A and C to B, $(AC)^2 + (BC)^2$, is maximum? With no loss in generality we can draw the geometry of this problem as shown in figure 1.1, with A on the y-axis. The figure shows A and B on the same side of the line, and places C between A and B, but as the analysis continues you'll see that these assumptions in no way affect the result.

In the notation of the figure we are to find the value of x that, with a, b, and c constants, maximizes $d_1^2 + d_2^2$. Now,

$$d_1^2 + d_2^2 = \{x^2 + a^2\} + \{(b - x)^2 + c^2\}$$
$$= a^2 + b^2 + c^2 + 2x\,(x - b).$$

Thus, we need to maximize the product $x(x - b)$; but we already know from Euclid how to do that—set $x = \frac{1}{2}b$. That is, C is midway between A and B. If you redraw figure 1.1 so that either $x > b$ or $x < 0$, and then write the expression for $d_1^2 + d_2^2$, you'll see that the result is unchanged.

An elementary example of an extremal problem in which there is (by the very nature of the problem) nothing to differentiate comes from discrete probability theory. Then the independent variable does not vary continuously but, rather, in discontinuous jumps. In such cases, taking a derivative simply has no meaning. So, suppose we toss four fair die, i.e., each one of the six faces on each die has probability $\frac{1}{6}$ of showing. What is the most likely number of die that will show a 3? The answer can only be one of five numbers, of course, the integers zero through four. If we define the value of the random variable X as the number of die that show a 3, then elementary probability theory tells us that $P(X = k)$ = probability that $X = k$ is given by

$$P(X = k) = \binom{n}{k}\left(\frac{1}{6}\right)^k\left(\frac{5}{6}\right)^{n-k},$$

where n is the number of die and $\binom{n}{k} = n!/(k!(n-k)!)$. So, with $n = 4$,

$$P(X = 0) = \frac{625}{1296}$$

$$P(X = 1) = \frac{500}{1296}$$

$$P(X = 2) = \frac{150}{1296}$$

$$P(X = 3) = \frac{20}{1296}$$

$$P(X = 4) = \frac{1}{1296}.$$

Thus, the most likely number of 3's to show is zero. But even more likely to happen is that *at least one* 3 shows, as

$$P(X \geq 1) = \sum_{k=1}^{4} P(X = k) = \frac{671}{1296} > P(X = 0).$$

This strikes many as a paradoxical result, but that is part of the inexhaustible charm of probability!

1.4 A Civil Engineering Problem

As a more sophisticated example of how minimization problems can sometimes be attacked with noncalculus approaches, consider the following. We have two towns, A and B, on opposite sides of a river

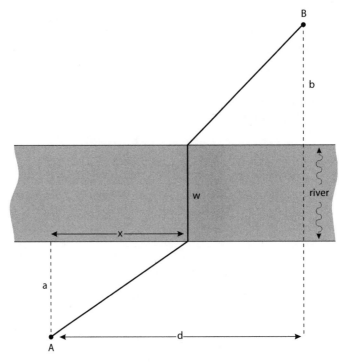

FIGURE 1.2. Minimum-distance bridge placement problem.

with constant width w. As shown in figure 1.2, A is distance a from the river, B is distance b, and the lateral separation of the two towns is d. Our problem is to determine where we should build a bridge over the river (perpendicular to the river's banks) so as to make the journey between A and B as short as possible. That is, what is x?

With calculus, this question is not hard to answer. We simply write the total distance as

$$T = \sqrt{a^2 + x^2} + w + \sqrt{b^2 + (d-x)^2}$$

and then set $dT/dx = 0$. Thus,

$$\frac{dT}{dx} = \frac{1}{2}\left[\frac{2x}{\sqrt{a^2+x^2}} - \frac{2(d-x)}{\sqrt{b^2+(d-x)^2}}\right]$$

and setting this equal to zero gives

$$x = \frac{ad}{a+b}.$$

Ancient mathematicians could also have solved this problem, however, long before the invention of the calculus, using just elementary geometry. To see how, let me first make a fundamental, exceedingly important and useful mathematical observation called the *triangle inequality*. The triangle inequality asserts that, given any triangle, the sum of any two of its sides is at least as large as the third side. It is really just a statement of the fact that the shortest path connecting two points in a plane is the straight line passing through the two points. Thinking of the triangle's sides as *directed* line segments with both magnitude and direction (i.e., as vectors), we can write $\vec{u}$ and $\vec{v}$ as two of the sides and $\vec{u} + \vec{v}$ as the third side, as shown in figure 1.3.

The triangle inequality says that $|\vec{u}| + |\vec{v}| \geq |\vec{u} + \vec{v}|$, where the absolute value signs denote the length of the vector. It is obvious that the inequality becomes an equality iff $\vec{u}$ and $\vec{v}$ point in the same direction (and so the triangle collapses to the "trivial triangle" with zero area). We can, in fact, now drop the imagery of the triangle itself, and simply think of $\vec{u}$ and $\vec{v}$ as *any* two vectors not necessarily associated with a triangle (although in many problems there will be a triangle).

Now, redraw figure 1.2 as figure 1.4 and label the various path segments as vectors. Notice that no matter what $\vec{x}$ is, the sum $(\vec{a} + \vec{x}) + (\vec{d} - \vec{x} + \vec{b})$ is constant. Mathematically this is trivial (the two $\vec{x}$'s

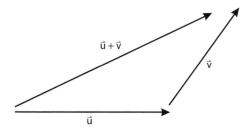

FIGURE 1.3. Vector addition.

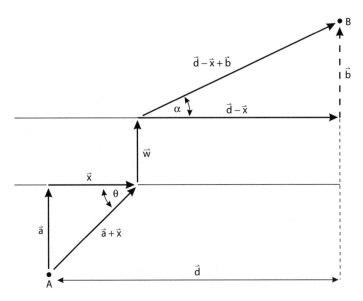

FIGURE 1.4. Bridge geometry in vector notation.

cancel), but physically this is because of the important observation that every vector sum (plus a constant $\vec{w}$ term to account for the bridge) starts at A and ends at B, no matter what $\vec{x}$ may be. By the triangle inequality $|\vec{a} + \vec{x}| + |\vec{d} - \vec{x} + \vec{b}| \geq |\vec{a} + \vec{d} + \vec{b}|$; an equality (which is the minimum sum) is achieved only when $\vec{a} + \vec{x}$ and $\vec{d} - \vec{x} + \vec{b}$ are in the same direction. That is, when $\theta = \alpha$ in the notation of figure 1.4.

Since the two triangles in figure 1.4 are right triangles with their other two angles equal, they are similar triangles. Thus, dropping the vector notation, we have

$$\frac{a}{x} = \frac{b}{d - x},$$

which is easily solved to give the location of the bridge at

$$x = \frac{ad}{a + b},$$

just as before. But this time no derivative was required. And, in fact, our ancient mathematician's solution actually provides some immediate extra physical insight that the calculus one does not; since $\theta = \alpha$, the path segments connecting each town to its respective river bank are *parallel*.

1.5 The AM-GM Inequality

There are yet other methods the mathematicians of old, in the days before calculus, could have used to solve many problems that seemingly require the calculation of derivatives. One of the most elegant of these methods is what is called the AM-GM inequality (the arithmetic mean-geometric mean inequality). It is easy to state:

> If $x_1, x_2, \cdots, x_n$ are any n positive numbers, $n \geq 1$, and
> if $A = (1/n)(x_1 + x_2 + \cdots + x_n)$ is the arithmetic mean of the x's
> and if $G = (x_1 x_2 \cdots x_n)^{1/n}$ is the geometric mean of the x's,
> then $A \geq G$ with equality iff $x_1 = x_2 = \cdots = x_n$.

New demonstrations of this famous and remarkably useful inequality appear on a regular basis to this day, but one of the easiest to understand (as well as one of the most elegant) is the 1954 proof by a mathematician named G. Ehlers. I know nothing more about Ehlers, but his proof of the AM-GM inequality is a gem and you can find it in appendix A. That proof uses just simple algebra and induction, but *no calculus*, which is appropriate since the whole point here is to show how we can solve many minimum/maximum problems without the techniques of calculus.

For example, recall the isoperimetric dual problem mentioned at the start of section 1.3: show that of all rectangles with a given area it is the square that has the smallest perimeter. This is actually quite easy to demonstrate with the AM-GM inequality. If we call the sides of the rectangle x and y, then the problem is to determine x and y so that we minimize

$$P = 2x + 2y = 2(x + y),$$

given that

$$A = xy$$

is a constant. From the AM-GM inequality with $n = 2$ we immediately have

$$\frac{1}{2}(x + y) \geq \sqrt{xy} = \sqrt{A}$$

with equality iff $x = y$. That is,

$$P = 2(x + y) \geq 4\sqrt{A},$$

which says P is never smaller than the *constant* $4\sqrt{A}$ and is equal to that constant iff $x = y$ (iff the rectangle is a square).

Closely related to this result is one concerning right triangles. Imagine all possible right triangles with perpendicular sides of lengths x and y that sum to a constant, i.e.,

$$x + y = k.$$

If we write A to denote the areas of the triangles, then

$$A = \frac{1}{2}xy.$$

Now, the AM-GM inequality for $n = 2$ says

$$\frac{x + y}{2} \geq \sqrt{xy} = \sqrt{2A}$$

with equality iff $x = y$. Thus,

$$\frac{k}{2} \geq \sqrt{2A},$$

or

$$A \leq \frac{k^2}{8}$$

with equality iff $x = y$. This shows that of all right triangles with perpendicular sides that sum to a constant, it is the *isosceles* right triangle that has the largest area (a result known since ancient times).

For another elegant illustration of the power of the AM-GM inequality, think back a bit to a question I asked you to ponder: of all triangles with a given area, show that it is the equilateral that has the smallest perimeter. Did you have any success doing that? It's not trivial! I'll do it here with the aid of the AM-GM inequality by showing the dual theorem: of all triangles with a given perimeter P, the equilateral has the largest area. As a prelude, recall the amazing formula for the area A of any triangle in terms of the lengths of its sides (a, b, and c). This formula is named after the Egyptian mathematician Heron of Alexandria, who is thought to have lived in the first century A.D. Some historians have speculated that the formula was known by Archimedes three centuries earlier, but there is no real evidence of that (other than Archimedes' genius, which makes it probable that he *did* know it), while the formula does appear in Heron's *Metrica*. It is not an easy formula to derive [see William Dunham, *Journey through Genius: The Great Theorems of Mathematics* (John Wiley 1990, pp. 118–27)], but it is easy to state:

$$A = \sqrt{s(s-a)(s-b)(s-c)},$$

where $s = \frac{1}{2}(a+b+c) = \frac{1}{2}P$, the so-called semiperimeter of the triangle. Since P is given, then so is s and Heron's formula tells us that to maximize A we must maximize the product $(s-a)(s-b)(s-c)$.

Notice first that each of the factors in that product is indeed positive, e.g.,

$$s - a = \frac{a+b+c}{2} - a = \frac{-a+b+c}{2} > 0$$

because from the triangle inequality for nontrivial triangles (triangles with nonzero area) we have $b + c > a$. Now, from the AM-GM inequality, we have

$$\frac{(s-a) + (s-b) + (s-c)}{3} = \frac{3s - (a+b+c)}{3} = \frac{3s - 2s}{3}$$

$$= \frac{s}{3} \geq [(s-a)(s-b)(s-c)]^{1/3}$$

with equality iff $(s-a) = (s-b) = (s-c)$, i.e., iff $a = b = c$. The term $s/3$ is a *constant upper*-bound to the inequality and so the area is maximized if $a = b = c$, and that's the entire proof!

As a third example of the AM-GM inequality solving a problem ordinarily thought to require calculus, consider the following question that probably appears in every calculus textbook ever written. A food can (with both ends sealed, of course) with the given volume V is to have the shape of a right circular cylinder. What are the dimensions of the can (the radius r and the height h) so that the surface area is minimum? The "calculus way" to answer this is to write the surface area S and the volume as

$$S = 2\pi r^2 + 2\pi rh$$
$$V = \pi r^2 h$$

and then to eliminate h. Thus, $h = V/\pi r^2$, and so

$$S = 2\pi r^2 + 2\pi r \frac{V}{\pi r^2} = 2\pi r^2 + \frac{2V}{r}.$$

We minimize S (as we'll see in chapter 4) by setting dS/dr to zero, i.e.,

$$\frac{dS}{dr} = 4\pi r - \frac{2V}{r^2} = 0,$$

which gives the solution for r. Thus, $V = 2\pi r^3$, or

$$\frac{V}{\pi r^2} = h = 2r.$$

That is, the height of the can with minimum surface area is equal to the diameter of the can.

Here's how the AM-GM inequality answers the same question. As before,

$$S = 2\pi \left(r^2 + rh\right) = 2\pi \left(r^2 + \frac{V}{\pi r}\right) = 2\pi \left(r^2 + \frac{V}{2\pi r} + \frac{V}{2\pi r}\right).$$

Or

$$\frac{S}{6\pi} = \frac{1}{3}\left(r^2 + \frac{V}{2\pi r} + \frac{V}{2\pi r}\right).$$

From the AM-GM inequality, we have

$$\frac{1}{3}\left(r^2 + \frac{V}{2\pi r} + \frac{V}{2\pi r}\right) \geq \left(r^2 \cdot \frac{V}{2\pi r} \cdot \frac{V}{2\pi r}\right)^{1/3} = \left(\frac{V^2}{4\pi^2}\right)^{1/3},$$

and so

$$\frac{S}{6\pi} \geq \left(\frac{V^2}{4\pi^2}\right)^{1/3} \quad \text{or} \quad S \geq 6\pi \left(\frac{V^2}{4\pi^2}\right)^{1/3}.$$

Thus, the surface area is never less than the *constant* $6\pi \left(\frac{V^2}{4\pi^2}\right)^{1/3}$, and is equal to that minimum value when $r^2 = \frac{V}{2\pi r} = \frac{V}{2\pi r}$, i.e., when $V = 2\pi r^3$ just as we found before (but before we had to know how to calculate a derivative).

Now, here's a little variation for you to play with: in the example just done, both ends of the can were sealed. Suppose instead that only the bottom end is sealed. For the same volume as before, what now is the relationship between r and h to minimize the surface area, and what is the ratio of the new minimized surface area to the one just calculated? It should be obvious that the ratio is less than one, but *how much* less than one? Remember, *no calculus!* There are *two* ways for you to attack this problem. You can start over and use the AM-GM inequality, of course. More clever, however, is to use our previous result, by noticing that if we take two cans, each with only one end sealed, and butt the unsealed ends together, we get a can with *both* ends sealed! Either way, you should get the same answers. (The answers are at the end of this section.)

We can use the AM-GM inequality to prove the following curious, and I think unobvious, fact: given two food cans of equal volume *and equal height*, one cylindrical and the other rectangular in shape, the cylindrical can will *always* have the smaller total surface area. To see this, observe that if V is the common volume, then, for either shape, we can write

$$V = (\text{area of bottom}) \times (\text{height}).$$

So, since the heights are also equal, then the areas of the bottoms (and tops) of the two shapes are equal, too. Thus, to decide which can shape has the smaller total surface area we need only to compare the vertical surface areas. To do that, let's make the following definitions:

$S_c =$ vertical surface area of a cylindrical can of radius r and height h, i.e., $S_c = 2\pi r h$,

$S_r =$ vertical surface area of a rectangular can with dimensions $a \times b \times h$, i.e., $S_r = 2ha + 2hb = 2h(a+b)$.

This means

$$S_r - S_c = 2h(a + b) - 2\pi rh = 2h[(a + b) - \pi r].$$

From the AM-GM inequality we have $(a + b) \geq 2\sqrt{ab}$, and so

$$S_r - S_c \geq 2h\left[2\sqrt{ab} - \pi r\right]$$

because I've replaced $(a + b)$ with a smaller quantity. Now, since the volumes of the two cans are equal we can also write

$$V = \pi r^2 h = abh,$$

and so

$$\sqrt{ab} = \sqrt{\frac{V}{h}}$$

and

$$\pi r = \pi\sqrt{\frac{V}{\pi h}} = \sqrt{\pi}\sqrt{\frac{V}{h}}.$$

This gives us

$$S_r - S_c \geq 2h\left[2\sqrt{\frac{V}{h}} - \sqrt{\pi}\sqrt{\frac{V}{h}}\right] = 2h\sqrt{\frac{V}{h}}\left[2 - \sqrt{\pi}\right] > 0$$

because it is clear that $2 > \sqrt{\pi}$ (i.e., $4 > \pi$). So, no matter how you choose the various dimensions of the two cans, if they have equal volume *and* equal height then the cylindrical can will *always* have the smaller total surface area.

If we don't require the two can shapes to have the same height, then it is no longer true that the cylindrical can will have the smaller surface area no matter what the dimensions may be. For example, suppose the rectangular can has dimensions $1 \times 1 \times \pi$, for a volume of π. Its total surface area is then $2 + 4\pi = 14.57$. If the cylindrical can has a radius of r and height h, then for the same volume we have $\pi r^2 h = \pi$, or

$$h = \frac{1}{r^2}.$$

Its total surface area is

$$T = 2\pi r^2 + 2\pi r h = 2\pi r^2 + 2\pi r \frac{1}{r^2}$$

$$= 2\pi \left(r^2 + \frac{1}{r} \right).$$

It is clear that we could pick r to make T arbitrarily larger than $2+4\pi$.

But it is also true that, if we pick r to give the minimum surface area for the cylindrical can, that area *will* be smaller than $2 + 4\pi$. That is, differentiating T gives

$$\frac{dT}{dr} = 2\pi \left(2r - \frac{1}{r^2} \right)$$

which is zero when $r = \left(\frac{1}{2} \right)^{1/3}$, which gives

$$T = 2\pi \left[\left(\frac{1}{2} \right)^{2/3} + \frac{1}{\left(\frac{1}{2} \right)^{1/3}} \right] = 2\pi \frac{\frac{1}{2} + 1}{\left(\frac{1}{2} \right)^{1/3}} = 2\pi 2^{1/3} \cdot \frac{3}{2}$$

$$= 3\pi 2^{1/3} = 11.87,$$

nearly 19% less than the surface area of the rectangular can.

As the final example of this section, let me show you how mathematicians of old could have solved yet another maximum problem. As shown in appendix B, using nothing but algebra (no calculus), a consequence of the AM-GM inequality is yet another inequality called the arithmetic mean-quadratic mean inequality (the AM-QM inequality): if $x_1, x_2, \cdots, x_n$ are n numbers, then

$$\frac{x_1 + x_2 + \cdots + n_n}{n} \leq \sqrt{\frac{x_1^2 + x_2^2 + \cdots + x_n^2}{n}}, \qquad n \geq 1$$

with equality iff $x_1 = x_2 = \cdots = x_n$. But the AM-GM inequality itself tells us that

$$(x_1 x_2 \cdots x_n)^{1/n} \leq \frac{x_1 + x_2 + \cdots + x_n}{n}$$

with equality iff $x_1 = x_2 = \cdots = x_n$, and so

$$(x_1 x_2 \cdots x_n)^{1/n} \leq \sqrt{\frac{x_1^2 + x_2^2 + \cdots + x_n^2}{n}}$$

with equality iff $x_1 = x_2 = \cdots = x_n$.

This general result has a very pretty geometric interpretation for $n = 2$, i.e., for

$$\sqrt{x_1 x_2} \leq \sqrt{\frac{x_1^2 + x_2^2}{2}}.$$

Suppose that $x_1^2 + x_2^2 = R^2$ (a constant). The equation $x_1^2 + x_2^2 = R^2$ is a circle (centered on the origin of the x_1, x_2 coordinate system) with radius R, and so $\sqrt{x_1 x_2}$ is bounded from above by the constant $R/\sqrt{2}$. And since $4x_1 x_2$ is the area of a rectangle inscribed in that circle, then that area is bounded from above by the constant $2R^2$ and that area is equal to $2R^2$ iff $x_1 = x_2$. That is, the inscribed rectangle of maximum area is the inscribed square.

The answers to the problem of the cylindrical can with minimum surface area, with just one end sealed, are

a. $r = h$

b. ratio of surface areas $= \dfrac{1}{2} \sqrt[3]{4} = 0.7937.$

1.6 Derivatives from Physics

There are minimum/maximum problems of great interest that *do* contain derivatives, but *not* because we are going to set them equal to zero. They are present because, for example, the *physics* of the problem requires them. The actual determination of a minimum (or a maximum) of something in such problems, however, depends on other sorts of arguments. So, for the penultimate section of this introductory chapter, let me take you through the details of one such problem that has derivatives aplenty because of the *physics* and not because of the mathematics.

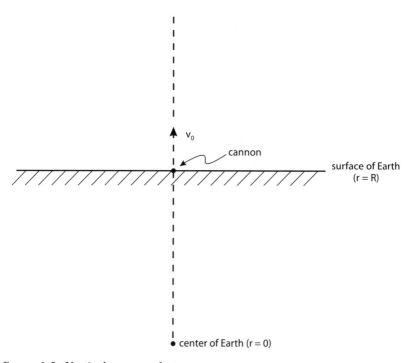

FIGURE 1.5. Vertical cannon shot.

Consider figure 1.5. There we have a cannon pointing straight up, directly away from the center of the earth (not drawn to scale!). If we fire the cannon a shell is ejected with initial velocity v_0, it rises upward to some maximum height, stops, and then falls back down to the ground. It is clear that the larger v_0, the higher the shell will go before gravity brings its upward motion to a halt. We can show, in fact, that if v_0 has a certain critical minimum value, then the shell will *not* return to earth. That minimum value for v_0 is called the *escape* velocity.

If we measure distance from the center of the earth as r ($r = 0$ *is* the center, and $r = R$ is the surface of the earth), then Newton's second law of motion (force equals mass times acceleration) and his inverse-square law of gravity tells us that if we ignore air-drag on the shell, then

$$m\frac{d^2r}{dt^2} = -G\frac{Mm}{r^2}, \qquad r \geq R,$$

where: m = mass of the shell,
         M = mass of the earth,
         G = universal constant of gravitation.

The minus sign on the right side of the differential equation is present because increasing r is directed upward, while the gravitational force on the shell is in the opposite direction, downward toward the center of the earth.

We can solve this second-order differential equation with the help of a powerful result from differential calculus called the chain rule (discussed in chapter 4): if we write $v(r)$ as the velocity of the shell at distance r from the center of the earth, then by definition

$$v = \frac{dr}{dt},$$

and so the acceleration of the shell is

$$\frac{d^2r}{dt^2} = \frac{dv}{dt} = \frac{dr}{dt} \cdot \frac{dv}{dr} = v\frac{dv}{dr}.$$

This reduces our original differential equation to the more tractable (with m canceled on both sides) equation

$$v\frac{dv}{dr} = -GM\frac{1}{r^2}, \qquad r \geq R.$$

We can "separate the variables" in this equation and write

$$v \, dv = -GM\frac{dr}{r^2},$$

which is easily integrated to give

$$\frac{1}{2}v^2 = GM\frac{1}{r} + C,$$

where C is the so-called "constant of indefinite integration." Now, since $v = v_0$ when $r = R$, then

$$\frac{1}{2}v_0^2 = GM\frac{1}{R} + C,$$

or

$$C = \frac{1}{2} v_0^2 - GM \frac{1}{R},$$

and thus

$$\frac{1}{2} v^2 = GM \frac{1}{r} + \frac{1}{2} v_0^2 - GM \frac{1}{R}.$$

If we define H as the shell's maximum distance from the center of the earth, then, as by definition $v = 0$ when $r = H$, we have

$$0 = \frac{GM}{H} + \frac{1}{2} v_0^2 - \frac{GM}{R},$$

or

$$H = \frac{GM}{\dfrac{GM}{R} - \dfrac{1}{2} v_0^2}.$$

If $v_0 = 0$ then $H = R$, which is simply the obvious; if the shell "leaves" the cannon with zero initial velocity, then it doesn't go anywhere! But as v_0 increases from zero, then H increases from R and, obviously, as $\frac{1}{2} v_0^2$ approaches GM/R we see that H diverges to infinity, i.e., the shell does not return to earth. So, the minimum escape velocity is the initial velocity given by

$$v_0 = \sqrt{\frac{2GM}{R}}.$$

Any velocity greater than this also means the shell isn't coming back, of course.

We can express this result in the following interesting alternative way. When $r = R$, the gravitational force on the shell is simply what we call its weight at the surface of the earth, which is mg, where g is the acceleration of gravity *at the surface*. Thus,

$$mg = G \frac{Mm}{R^2},$$

and so $GM = g R^2$. This gives the escape velocity as

$$v_0 = \sqrt{\frac{2g R^2}{R}} = \sqrt{2g R}.$$

Taking the earth's radius as 3,950 miles, and g as 32.2 ft/sec^2, we have the escape velocity as

$$v_0 = \sqrt{2 \times 32.2 \times 3{,}950 \times 5{,}280} \text{ ft/sec}$$

$$= 36{,}649 \text{ ft/sec} = 6.94 \text{ miles/sec}.$$

This is not the way we send people into space, of course, as the initial acceleration of the shell (spaceship) from zero to almost seven miles per second over the length of a cannon barrel would be unsurvivable. (But see Jules Verne's *From the Earth to the Moon*. In his 1865 novel, he proposed getting around the problem of shooting men to the moon using a fantastic 900-foot-long cannon. It wouldn't work, but it *is* clever.) But, serious proposals *have* been made to put nonhuman payloads into orbit or on the moon, using super-high acceleration up to the escape velocity. Such accelerations would be achieved not with a cannon but, rather, with the far more exotic technology of electromagnetic launchers, which are in actual use today at several sophisticated rollercoaster rides around the world.

1.7 Minimizing with a Computer

For the final two examples of this chapter, which return to the theme of the computer as a useful tool in extremal problems, suppose first that a man can walk n times faster than he can swim (it seems reasonable that $n \geq 1$, but I'll not use that assumption in what follows). He wants to travel from A, on the edge of a circular lake with radius R (centered on point O) to C, also on the edge of the lake. C's location is specified by the given angle β (measured from the diameter AOD), as shown in figure 1.6. His general strategy is to first swim along the chord AB, and then to walk the rest of the way along the lake's edge from B to C. If his total travel time is T, then where should B be to minimize T?

If we denote by θ the central angle subtended by the man's walk, then the isosceles triangle OAB (with the chord AB as its base) has equal base angles of α and a third angle of $\gamma = \pi - \theta - \beta$. Thus,

$$(2\alpha) + (\pi - \theta - \beta) = \pi \text{ radians,}$$

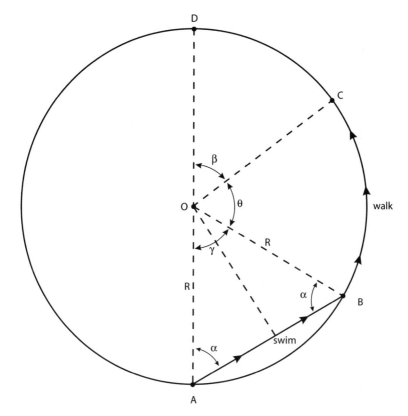

FIGURE 1.6. Crossing a circular lake in minimum time.

or

$$\alpha = \frac{1}{2}\left(\theta + \beta\right).$$

It is clear from figure 1.6 that the man's swimming and walking distances are, respectively, $2R\cos\left\{\frac{1}{2}\left(\theta + \beta\right)\right\}$ and $R\theta$. So, if we call his swimming speed unity (in arbitrary units) then his walking speed is n and we have the total travel time as

$$T = 2R\cos\left\{\frac{1}{2}\left(\theta + \beta\right)\right\} + \frac{R\theta}{n}$$

$$= R\left[2\cos\left\{\frac{1}{2}\left(\theta + \beta\right)\right\} + \frac{\theta}{n}\right].$$

As a quick, partial check on this expression, notice that if $\beta = \pi$ radians ($C = A$) then we also have $\theta = 0$ and $T = 0$, just as we should have (it doesn't take any time to travel from where you are to where you are!).

Our problem then is simply this: given a value of β in the interval 0 to π (thus locating C), what θ minimizes T (thus locating B)? This is an easy question to study with the aid of a computer. Figure 1.7 shows how T varies with θ, for five values of n, with $\beta = 0$ (C is directly across the lake from A) and figure 1.8 assumes $\beta = 90°$. In both figures the constant scale factor of R in the expression for T has been ignored since it has no affect on the value for θ that gives an *extrema* in T.

The plots in the two figures contain a wealth of information. In figure 1.7, for example, the $n = 1$ and $n = 1.5$ curves have their minimum values at $\theta = 0$ (the man should *swim*, all the way, from A to C), while the $n = 2$, $n = 2.5$, and $n = 3$ curves have their

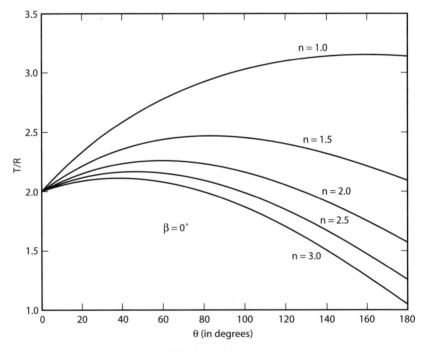

FIGURE 1.7. Total travel time across the lake, $\beta = 0°$.

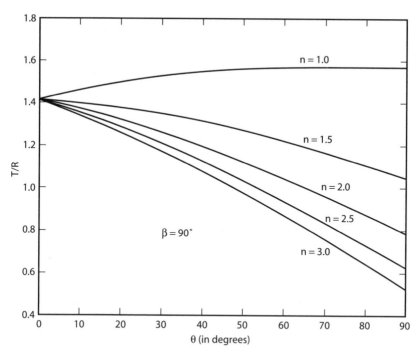

FIGURE 1.8. Total travel time across the lake, $\beta = 90°$.

minimum values at $\theta = 180°$ (the man should *walk*, all the way, from
A to C). The curves suggest that there is some value of n between 1.5
and 2 where *either* of the pure walk-only and swim-only strategies
would give the minimum travel time. What is that critical value of
n? A little thought should convince you it is $n = \frac{1}{2}\pi = 1.57$. The
curves of figure 1.8 suggest the same general conclusion for $\beta > 0$,
i.e., as n increases from unity the strategy for minimizing the total
travel time begins as the pure strategy of swimming all the way and
then switches to the pure strategy of walking all the way. Is this
always true? That is, for any value of β, is it true that there is never
a mixed strategy of walking *and* swimming that minimizes T? I'll
leave that for you to think about!

For my last example in this chapter, consider the following problem that is superficially similar to the one just treated, but which
offers some surprising complications. But *not* so much complication that we can no longer make a fruitful computer analysis. So,

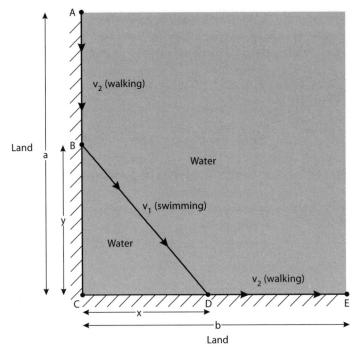

FIGURE 1.9. Another water-crossing problem.

suppose now that the man is initially at point A on a beach with a right-angle bend, as shown in figure 1.9. The man wishes to travel from A to E in minimum time; at any point B, as he walks along the first section of beach toward C, he can enter the water and swim to D, where he exits the water and continues walking on the second section of beach to E. That is, he can "cut a corner" from one section of beach to the other. The lengths of the two sections of beach are a and b, as shown in figure 1.9.

It is not difficult to express the problem mathematically. If we write v_1 and v_2 for the man's speeds while swimming and walking, respectively, and if x and y are the distances of points D and B from the corner of the beach (C), respectively, then the total travel time is a function of *two* variables:

$$T(x, y) = \frac{a - y}{v_2} + \frac{\sqrt{x^2 + y^2}}{v_1} + \frac{b - x}{v_2}$$

$$= \frac{(a+b)-(x+y)}{v_2} + \frac{\sqrt{x^2+y^2}}{v_1}.$$

Our problem, then, is to determine the values of x and y that minimize T for given values of a, b, v_1, and v_2.

The answer for $v_1 > v_2$, for *any* a and b, is physically obvious: $x = b$ and $y = a$, i.e., the man *swims* the entire trip because then he travels the straight line path (shortest possible path) from A to E at the greater speed. As argued before, swimming faster than he can walk isn't very plausible, however, and the case of $v_1 < v_2$ is far more interesting (both physically *and* mathematically). Before continuing with the analysis of $T(x, y)$, it is important to notice that, with a *single* exception, the values of x and y are independent, subject only to the constraints of $0 \le x \le b$, $0 \le y \le a$. The single exception is that if either x or y is zero then so must be the other; this is because of the physically required *continuous* nature of a path from A to E.

Now, we *could* attack the problem of minimizing $T(x, y)$ with the aid of rather sophisticated calculus, but that isn't attractive for several reasons. First, that would be out of place so early in this book and, second, there is a very pretty *geometric* interpretation of the problem. Indeed, you'll see the same approach used later, when we get to linear programming in chapter 7. And third, the approach I'll show you now makes great use of the sheer computational power of a computer.

To begin, all pairs of points (x, y) that satisfy the constraints $0 \le x \le b$, $0 \le y \le a$ form what is called the set of *feasible* solutions. For our problem, this set is the rectangle shown in figure 1.10, with the understanding that the bottom edge ($x > 0$, $y = 0$) and the left vertical edge ($x = 0$, $y > 0$) are *not* included in the feasible solution set; the corner *point* $(0, 0)$ *is*, however, in the feasible solution set. We want to find the point in the feasible solution set that minimizes $T(x, y)$. Now, notice that we can write

$$v_1 v_2 T = v_1(a+b) - v_1(x+y) + v_2\sqrt{x^2+y^2},$$

or

$$\sqrt{x^2+y^2} - \left(\frac{v_1}{v_2}\right)(x+y) = U,$$

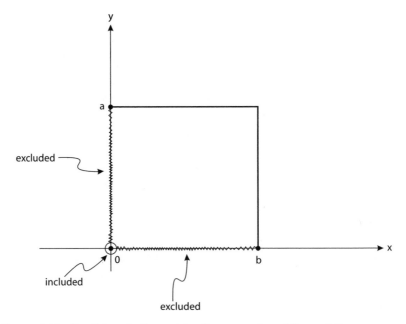

FIGURE 1.10. Feasible solution set for the geometry of figure 1.7.

where

$$U = v_1 T - \left(\frac{v_1}{v_2}\right)(a + b).$$

Since v_1, v_2, a, and b are given positive constants, then it is clear that the minimization of T is equivalent to the minimization of U. This simple observation turns out to be the key observation in the following analysis.

The equation

$$\sqrt{x^2 + y^2} = \left(\frac{v_1}{v_2}\right)(x + y) + U$$

defines a curve $y = y(x)$ for any given U; as we vary U we will also vary the curve $y = y(x)$. We wish to determine the minimum U that results in a curve that still passes through at least one point of the feasible solution set. Using a computer to draw these curves will give us all the insight we need to determine the minimizing $U(= U_{min})$ and, hence, the minimized $T(= T_{min})$:

$$T_{\min} = \frac{1}{v_2}(a+b) + \frac{1}{v_1}U_{\min}.$$

To plot $y = y(x)$ as a function of U, it is convenient to change to polar coordinates:

$$x = r\cos(\theta)$$
$$y = r\sin(\theta),$$

and so

$$\sqrt{r^2\cos^2(\theta) + r^2\sin^2(\theta)} = \left(\frac{v_1}{v_2}\right)[r\cos(\theta) + r\sin(\theta)] + U.$$

This is easily reduced to

$$r = \frac{U}{1 - \left(\dfrac{v_1}{v_2}\right)[\sin(\theta) + \cos(\theta)]},$$

where, of course, it is understood that the radius vector r (at polar angle θ from the origin to the arbitrary point (x, y) on the $y(x)$ curve) is always nonnegative, i.e., $r \geq 0$. That is, the numerator and the denominator must have the same sign.

For the remainder of this analysis, let's assume that both the numerator and denominator are nonnegative, i.e., that

$$U \geq 0$$

$$1 - \left(\frac{v_1}{v_2}\right)[\sin(\theta) + \cos(\theta)] \geq 0.$$

Since $f(\theta) = \sin(\theta) + \cos(\theta)$ achieves a maximum value of $\sqrt{2}$ at $\theta = 45°$ (easily verified by either setting $df/d\theta = 0$ or by simply plotting $f(\theta)$), then as long as

$$\left(\frac{v_1}{v_2}\right) \leq \frac{1}{\sqrt{2}},$$

we will have $r \geq 0$ for any $U \geq 0$ for all values of the polar angle θ. That is, we are now dealing with a restrictive case of $v_1 < v_2$, i.e., with $v_1 \leq (1/\sqrt{2})v_2$.

Returning to the original x, y coordinate system, we have the result we are after: the $y = y(x)$ curve is the curve defined by

$$x = \frac{U\cos(\theta)}{1 - \left(\dfrac{v_1}{v_2}\right)[\sin(\theta) + \cos(\theta)]},$$

$$y = \frac{U\sin(\theta)}{1 - \left(\dfrac{v_1}{v_2}\right)[\sin(\theta) + \cos(\theta)]}.$$

We can see now that all U "does" is *scale* the plot. Indeed, in figure 1.11 you'll find the curve $y = y(x)$ for $v_2 = 5$ with four different values of v_1 (all satisfy the condition $v_1 \leq (1/\sqrt{2})v_2$), for two values of U (solid for $U = 1$, dashes for $U = 0.4$). It is clear from these plots that $y = y(x)$ is elliptical, and that as U decreases toward zero, the curves shrink inward to around the lower-left-corner point of the feasible solution set (in solid).

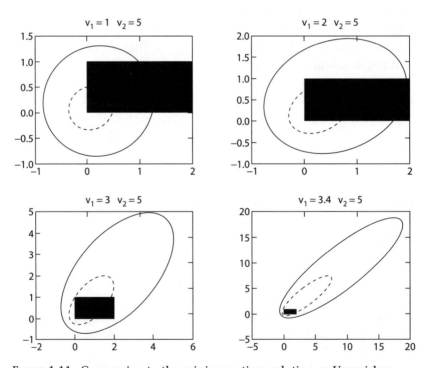

FIGURE 1.11. Converging to the minimum-time solution as U vanishes.

I have, to be specific, used $a = 1$ and $b = 2$ to define the feasible solution rectangle, but it should be obvious that the actual size of the rectangle doesn't matter. That is, for any choice of the a and b values, the smallest nonnegative U is $U = 0$, which collapses the $y = y(x)$ curve to the single point $x = 0$, $y = 0$. Since the smallest U gives the smallest T, then $T(0, 0)$ is the minimum journey time.

Thus, somewhat surprisingly I think, if $v_1 \leq (1/\sqrt{2})v_2$ then the man should *walk* the *entire* way, and that is so no matter what are the dimensions of the beach. So, we have solved the problem for the two cases of $v_1 \geq v_2$ (swim all the way) and $v_1 \leq (1/\sqrt{2})v_2$ (walk all the way). What if $(1/\sqrt{2})v_2 < v_1 < v_2$? I'll leave that case for *you* to ponder!

How to Walk Out of the Woods

Our lost hiker doesn't know which way to go to walk directly back to his car, but he does know that the car is *somewhere* on the circumference of the circle, with a one-mile radius, centered on his present location. So, to insure he returns to his car, he should first walk one mile in a randomly selected direction—if he is *very* lucky he'll walk straight back along the radius that was his original path—and then walk along the circular (one-mile-radius) path centered on his starting point. *Somewhere* along that circular path is his car. The absolute maximum distance he'll have to walk is the initial one-mile radius plus the 2π-mile circumference, i.e., $1 + 2\pi = 7.2832$ miles.

This is a mathematician's solution, of course, as it ignores the practical detail of just *how* one manages to walk along a circular arc in a densely wooded forest. Another setting for this problem, that avoids that objection, is to have our lost soul be a fisherman in a rowboat one mile off shore, in a dense fog. *Rowing* in a circle is now "easy"; all the fisherman need do is to take one end of a rope, drop it overboard with a heavy anchor, measure the depth of the water, and then (with due regard for the depth) row away until enough rope has played out to put him a mile away. He can then, keeping the rope taunt, swing in a circular path about his original position.

Now, here's a new twist on this puzzle for you to think about. Is this solution the best one can do, where *best* means

(continued)

having the minimum maximum path length? The answer is *no*, there are paths that require smaller maximum travel distances that, with certainty, return the fisherman back to shore (this is not quite the same as getting back to the *car* itself, of course, but for both the hiker and fisherman it is probably good enough!)

To see this, imagine our lost fisherman first picks some angle $\theta > 0$, and then at random picks a direction that he assumes is the direct one-mile path to the shore. He then rows at angle θ to this line for a distance of $\sqrt{1 + \tan^2(\theta)}$, as shown in figure

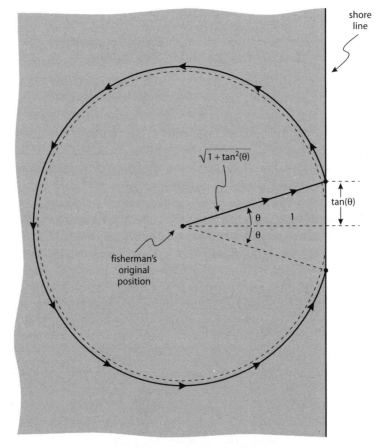

FIGURE 1.12. Geometry of the lost fisherman problem.

(continued)

1.12. That is, the triangle formed by his initial position, his new position, and the end of the one-mile path in the *assumed* direction to the shore, is a right triangle. If the assumed direction to the shore happens to be correct, then his journey is over. Otherwise, he next rows along a circular path with radius $\sqrt{1 + \tan^2(\theta)}$ until the line from his original position to his present position is once again θ with respect to the assumed one-mile path. That is, he rows along a circular path through an angle of $2\pi - 2\theta$ radians. Since the original solution was sure to eventually return him to shore, it is clear from figure 1.12 that this new path will also eventually reach the shore as well (since the original solution path lies entirely *inside* the new path). The maximum total length of this new path is

$$L(\theta) = \sqrt{1 + \tan^2(\theta)} + 2\pi \sqrt{1 + \tan^2(\theta)} \left(\frac{2\pi - 2\theta}{2\pi} \right)$$

$$= [1 + 2\pi - 2\theta] \sqrt{1 + \tan^2(\theta)}.$$

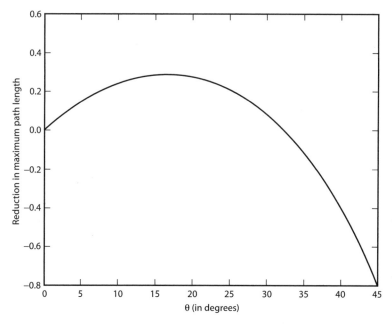

FIGURE 1.13. Proof that $2\pi + 1$ is not the minimum path length.

(continued)

Notice that $L(0) = 1+2\pi$, the maximum length of the original solution. The astonishing result is that there are values for $\theta > 0$ that *do* result in $L(\theta) < L(0)$!

This claim is easily established by simply plotting the quantity $L(0)-L(\theta)$ versus θ, as shown in figure 1.13. (We might try setting the derivative of $L(\theta)$ to zero, of course, to calculate the value of θ that minimizes $L(\theta)$, but if you do that you'll find you are led to a transcendental equation in θ, i.e., you will *still* need to use a computer—see section 4.5.) Figure 1.13 shows that, for $\theta = 16.61°$, $L(\theta)$ is 0.2879 miles *less* than 7.2832 miles.

Now, one final question for you—can our fisherman do *even better*? Is there a rowing path that has an even smaller maximum length that is still certain to get him to shore? The answer is again *yes*, and an analysis demonstrating that is given in appendix H—but don't look until you've made an honest try.

2.

The First Extremal Problems

2.1 The Ancient Confusion of Length and Area

Ancient mathematicians, the Greeks and the Egyptians of the several centuries before Christ, treated a number of questions of the type we are interested in. They included the *isoperimetric problem* (what closed curve of given length encloses the greatest area?), and such questions as how to determine the line of minimum length that joins a given point to a given curve. Apollonius of Perga (262–190 B.C.) gave many ingenious geometric constructions to the latter question in his work *Conics*, but generally such problems are now handled easily with calculus. I'll not discuss Apollonius' solutions here, then, but if you are curious, you can see how he reasoned in volume 2 of Thomas Heath's classic work *A History of Greek Mathematics*, (Oxford 1921, pp. 159–63).

At just about the same time, the great Archimedes (287–212 B.C.) had tackled a fascinating problem concerning the volumes of the spherical caps cut off by planes passing through spheres of various radii, with the constraint that the caps all have the same surface area. (A *spherical cap* is the region of a sphere that lies above, or below, a plane that cuts through a sphere. Any plane passing through a sphere's center, for example, divides the sphere into two equal spherical caps called *hemispheres*.) In his masterpiece *De Sphaera et Cylindro* (*On the Sphere and the Cylinder*), Archimedes showed that of all such equal-area caps it is the hemispherical cap that has the largest volume. Again, I won't go into Archimedes' geometric demonstration of this but, if curious, you can read how he did it in Thomas Heath's 1897 book *The Works of Archimedes* (reprinted in 1953 by Dover Publications), pp. 88–90.

Yet another extremal problem of ancient origin is the *geodesic problem in the plane* (what is the curve of minimum length that joins two given points?). The answer was intuitively accepted by the ancients as a straight line, and so will we *for the time being*. In fact, however, the isoperimetric and the geodesic questions, all too easily dismissed by students as having "obvious" answers, are actually extremely deep questions that stretched brilliant minds to find and *prove* the answers. The answer to the first (a circle) defied a rigorous derivation until the nineteenth century (!), while the answer to the second was formally proven only just a bit earlier (in the eighteenth century). The ancients "knew" the answers long before these modern proofs, of course, and their proofs are actually quite convincing. But they contain a common, very subtle flaw (by modern standards), the explanation of which I'll save for the last section of this chapter.

To say that the ancients knew the answer to the isoperimetric problem, however, is not to say it was commonly known. There is, for example, an amusing passage in Book 4 of Polybius' *Histories* (of the Greek world more than a century before Christ) that shows this. Titled "Computation of the size of Cities," it reads:

> Most people judge the size of cities simply from their circumference. So that when one says that Megalopolis is fifty stades in circumference [about five miles] and Sparta forty-eight, but that Sparta is twice as large as Megalopolis, the statement seems incredible to them. And when in order to puzzle them still more, one tells them that a city or camp with a circumference of forty stades may be twice as large as one the circumference of which is one hundred stades, this statement seems to them absolutely astounding. The reason of this is that we have forgotten the lessons we learnt as children. I was led to make these remarks by the fact that not only ordinary men but even some statesmen and commanders of armies are thus astounded, and wonder how it is possible for Sparta to be larger and even much larger than Megalopolis, although its circumference is smaller; or at other times attempt to estimate the number of men in a camp by taking into consideration its circumference alone. . . . So much for those who aspire to

political power and the command of armies but are ignorant of such things and surprised by them.

Polybius wrote that in the second century B.C., but the ignorance he was complaining about was difficult to overcome. Six hundred years later, for example, we find in a commentary written by the mathematically trained philosopher Proclus (on the first book of Euclid's *Elements*) the following warning about the possibility of being short-changed by someone who has not forgotten his geometry lessons:

> We often fail to watch out for [the error of equating area with perimeter] in the distribution of plots of land; and many persons have taken the larger of two plots and [improperly] got a reputation for justice as having chosen an equal portion because the sum of the boundaries is the same in both cases.

Proclus gives the further interesting example of two isosceles triangles, one with sides 5, 5, and 6, and the other with sides 5, 5, and 8. The unwary might assume the first to have the smaller area because it has the smaller perimeter, but in fact a quick application of Heron's area formula (or of the 3, 4, 5 right triangle geometry created by drawing the altitude to the longest side) shows that the two triangles have the same area (of 12).

With these words of Polybius and Proclus in mind, it is now easy to understand why the average Greek of ancient times found it paradoxical that the two triangles shown in figure 2.1 should have the same area (because they have the same base and equal height), even though the perimeter of *A* is clearly less than that of *B*. Indeed,

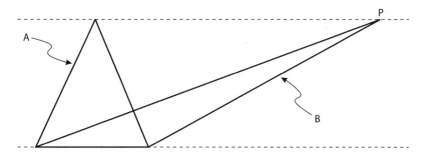

FIGURE 2.1. Two triangles with equal areas but unequal perimeters.

by sliding the vertex P of B arbitrarily far to either the left or right along the upper dashed line we can increase the perimeter of B without bound, without changing the area of B. This increase in perimeter comes with a price, of course: we need a bigger and bigger "expanse" of the plane to contain B even though its area remains constant. What is still astonishing to this day is that it is possible to do this example one better; to have a closed, simple (i.e., non-self-intersecting) curve that bounds finite area with an infinite perimeter in a *finite* region of the plane.

In 1906, for example, the Swedish mathematician Helge von Koch (1870–1924) published what has come to be called the "von Koch snowflake," a closed curve of infinite length that lies totally within a finite region of the plane (and so encloses a finite area). The iterative construction of this astonishing curve is easy to describe. We start with an equilateral triangle, with sides of unit length. Then, as the first iteration, the middle third of each side is removed and replaced with equilateral triangles with sides of length $\frac{1}{3}$. Then, as the second iteration, the middle third of the sides in the first iteration curve are removed and replaced with equilateral triangles with sides of length $\frac{1}{9}$. And so on indefinitely, as suggested in figure 2.2; the von Koch snowflake is the curve that results as the number of iterations increases without limit.

To see the astonishing perimeter/area property of the von Koch snowflake, let's make the following definitions. After the nth iteration, $n \geq 0$,

$$N_n = \text{number of sides}$$
$$\ell_n = \text{length of each side}$$
$$L_n = \text{length of perimeter} = N_n \ell_n.$$

So, $\ell_0 = 1$, $N_0 = 3$, and $L_0 = 3$. It is obvious that with each iteration the number of sides increases by a factor of 4 (inserting a triangle in the middle of a side increases *one* side to *four* sides—the original side is split into two sections, plus the two sides of the triangle itself). Since $N_0 = 3$, then

$$N_n = 3 \cdot 4^n, \qquad n = 0, 1, 2, \cdots.$$

Also obvious is that with each iteration the length of a side decreases by a factor of 3. Since $\ell_0 = 1$, then

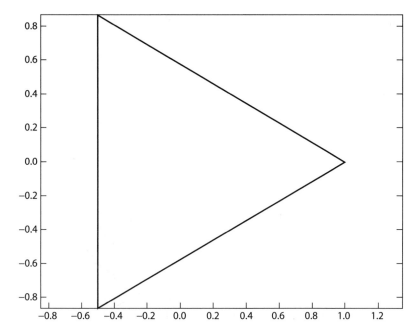

FIGURE 2.2a. von Koch snowflake iteration 0.

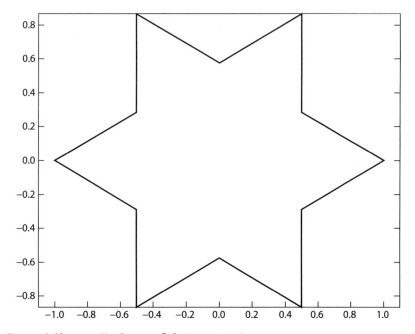

FIGURE 2.2b. von Koch snowflake iteration 1.

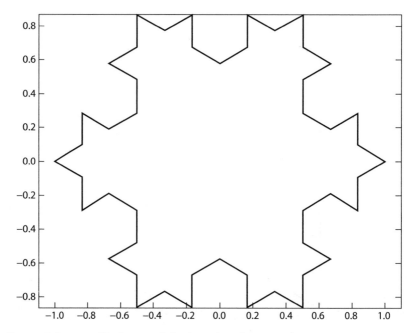

FIGURE 2.2c. von Koch snowflake iteration 2.

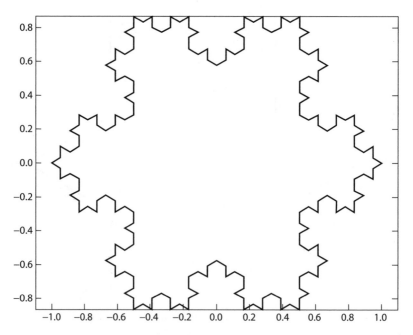

FIGURE 2.2d. von Koch snowflake iteration 3.

$$\ell_n = 1 \cdot \left(\frac{1}{3}\right)^n = \frac{1}{3^n}, \qquad n = 0, 1, 2, \cdots.$$

Thus, the perimeter after the nth iteration is

$$L_n = N_n \ell_n = 3 \cdot 4^n \cdot \frac{1}{3^n} = 3 \cdot \left(\frac{4}{3}\right)^n,$$

and so

$$\lim_{n \to \infty} L_n = \lim_{n \to \infty} 3 \cdot \left(\frac{4}{3}\right)^n = \infty.$$

It is probably obvious as well that the total area bounded remains finite as $n \to \infty$ because each iteration results in an increasingly "crinkly" curve (*so* crinkly, in fact, that in the limit $n \to \infty$ we can't draw it!) through the use of ever smaller triangles. Certainly all of the iterative curves remain inside a circle with a radius of, say, 1. But, to be sure this claim of finite area is clear, let's calculate the precise area of the von Koch snowflake.

To begin, observe that the area of an equilateral triangle with side lengths ℓ_n is, by Heron's formula from chapter 1, with $s = \frac{1}{2}$ $(\ell_n + \ell_n + \ell_n) = \frac{3}{2}\,\ell_n$, given by

$$\sqrt{s(s - \ell_n)(s - \ell_n)(s - \ell_n)} = \sqrt{\frac{3}{2}\,\ell_n \cdot \frac{1}{2}\,\ell_n \cdot \frac{1}{2}\,\ell_n \cdot \frac{1}{2}\,\ell_n} = \frac{\sqrt{3}}{4}\,\ell_n^2.$$

So, for example, if A_n is the area bounded after the nth iteration, then

$$A_0 = \frac{\sqrt{3}}{4}.$$

Now, with each iteration we clearly increase the enclosed area; from A_{n-1} we increase to A_n by adding $3 \cdot 4^{n-1}$ equilateral triangles (a triangle for each side) with $\ell_n = \frac{1}{3^n}$. (For example, with $n = 1$ we increase from A_0 to A_1 by adding three equilateral triangles, each with $\ell_1 = \frac{1}{3}$.) The area of each one of these added triangles is

$$\frac{\sqrt{3}}{4} \cdot \left(\frac{1}{3^n}\right)^2 = \frac{\sqrt{3}}{4} \cdot \frac{1}{9^n},$$

and so

$$A_n = A_{n-1} + 3 \cdot 4^{n-1} \cdot \frac{\sqrt{3}}{4} \cdot \frac{1}{9^n} = A_{n-1} + \frac{\sqrt{3}}{4} \left[\frac{3 \cdot 4^{n-1}}{9^n} \right]$$

$$= A_{n-1} + A_0 \frac{3 \cdot 4^{n-1}}{9 \cdot 9^{n-1}} = A_{n-1} + A_0 \cdot \frac{1}{3} \left(\frac{4}{9} \right)^{n-1}.$$

Writing this expression for A_n explicitly for the first few values of $n \geq 1$, we see that

$$n = 1: A_1 = A_0 + A_0 \frac{1}{3} \left(\frac{4}{9} \right)^0 = A_0 \left[1 + \frac{1}{3} \left(\frac{4}{9} \right)^0 \right];$$

$$n = 2: A_2 = A_1 + A_0 \frac{1}{3} \left(\frac{4}{9} \right) = A_0 \left[1 + \frac{1}{3} \left(\frac{4}{9} \right)^0 \right] + A_0 \frac{1}{3} \left(\frac{4}{9} \right),$$

$$\text{or} \quad A_2 = A_0 \left[1 + \frac{1}{3} \left(\frac{4}{9} \right)^0 + \frac{1}{3} \left(\frac{4}{9} \right) \right];$$

$$n = 3: A_3 = A_2 + A_0 \frac{1}{3} \left(\frac{4}{9} \right)^2 = A_0 \left[1 + \frac{1}{3} \left(\frac{4}{9} \right)^0 + \frac{1}{3} \left(\frac{4}{9} \right) \right]$$

$$+ A_0 \frac{1}{3} \left(\frac{4}{9} \right)^2,$$

$$\text{or} \quad A_3 = A_0 \left[1 + \frac{1}{3} \left(\frac{4}{9} \right)^0 + \frac{1}{3} \left(\frac{4}{9} \right) + \frac{1}{3} \left(\frac{4}{9} \right)^2 \right].$$

In general, then, we can write

$$\lim_{n \to \infty} A_n = A_0 \left[1 + \frac{1}{3} \left(\frac{4}{9} \right)^0 + \frac{1}{3} \left(\frac{4}{9} \right) + \frac{1}{3} \left(\frac{4}{9} \right)^2 + \cdots \right]$$

$$= A_0 \left[1 + \frac{1}{3} \left\{ 1 + \left(\frac{4}{9} \right) + \left(\frac{4}{9} \right)^2 + \cdots \right\} \right].$$

The expression in the braces is a geometric series, i.e.,

$$1 + x + x^2 + \cdots = \frac{1}{1-x}, \qquad |x| < 1$$

and, with $x = 4/9$ in the expression for $\lim_{n\to\infty} A_n$, we have

$$\lim_{n\to\infty} A_n = A_0 \left[1 + \frac{1}{3} \cdot \frac{1}{1 - \frac{4}{9}} \right] = \frac{8}{5} A_0.$$

So, in the limit $n \to \infty$, the von Koch process increases the initial enclosed area by just 60%, while increasing the initial perimeter to infinity. The von Koch snowflake occupies a finite region of the plane as well, unlike triangle B in figure 2.1. Even Polybius and Proclus, I think, would have been astonished by the area/perimeter properties of the von Koch snowflake.

2.2 Dido's Problem and the Isoperimetric Quotient

The origin of the isoperimetric problem can be traced back to the legendary story of the Phoenician queen Dido, told by Virgil in his *Aenid*. In that tale of events supposed to have taken place in the middle of the ninth century B.C., we read of Dido fleeing from her brother Pygmalion, who has murdered her husband. Escaping by sea, she finally lands in North Africa in what today is called the Bay of Tunis. There she comes to an agreement with the local inhabitants that she may buy all the land that can be bounded by the hide of a bull. The locals must have thought that to be a great joke, but Dido had the last laugh; she cut the hide into a great many long, narrow strips and attached them end-to-end. Then, using the seashore (given as straight) as part of the boundary, she laid out the hide-strip to enclose the maximum possible area, which she "knew" would be in the shape of a semicircle. Thus was founded, so goes the legend, both the ancient city of Carthage as well as the problem of Dido.

Carthage disappeared long ago (destroyed for the last time at the end of the seventh century A.D.), but the problem of Dido has remained one of the classics of mathematics: to find, among all possible curves of fixed length that connect two points on another given curve, the one curve that bounds the largest area. For the original problem of this type the given curve is a straight line (the seashore) and, from the assumption that the solution curve is semicircle, it is then "easy to see" that the solution curve to the isoperimetric problem is a circle. You'll see how all this works by the end of this chapter.

The "Problem of Dido" is also less well known as the "Problem of Hengist and Horsa." The name comes from two German brothers, mercenaries that British legend says were hired to squash a fifth-century-A.D. invasion by the Saxons that resulted from Rome's withdrawal of its occupying legions because Rome itself was under attack (thus leaving Britain vulnerable). As payment for their services the brothers asked "only" for all the land that could be bounded by the hide of an ox (of course, they then did as Dido, cutting the hide into many thin strips and forming a large circle). This legend is at least as bloody and deceitful as is Dido's, but perhaps with a bit more romance— the tale serves as the prologue to the story of Merlin and King Arthur. The isoperimetric legend, like worldwide flood legends, seems to be common to many civilizations across both geography and time.

We will generally not be very much interested here in the metaphysical musings of philosophers, but Aristotle's passage in Book 2 of his *De caelo* (*On the Heavens*) is provocative, where he argues for circular motion of the stars:

> . . . the revolution of the heaven is the measure of all motions, because it alone is continuous and unvarying and eternal, the measure in every class of things is the smallest member, and the shortest motion is the quickest, therefore the motion of the heaven must clearly be the quickest of all motions. *But the shortest path of those which return upon their starting-point is represented by the circumference of a circle* [my emphasis] and the quickest motion is that along the shortest path.

Did Aristotle write this because he knew the solution to the isoperimetric problem? It certainly would seem so.

The first mathematical attack on the isoperimetric problem is thought to have appeared in the work *On Isometric Figures* by the somewhat mysterious Greek mathematician Zenodorus. Very little of his life is known. Even when he lived is open to some debate, but most historians place him shortly after the time of Archimedes, i.e., in the second century B.C. Indeed, there is mention by the Greek mathematical-philosopher Simplicius in the sixth century A.D. (in his commentary of *De caelo*) of a proof by Archimedes of the isoperimetric theorem, but many historians today believe that may be an error. Writing so soon after Archimedes, we might expect that Zenodorus himself would have had something to say about his predecessor's work, but unfortunately *On Isometric Figures* has been lost to history, with our knowledge about its contents formed only by what later writers had to say. In particular, from the commentaries written by the fourth-century-A.D. Egyptians Theon of Alexandria (on Ptolemy's *Syntaxis mathematica*, better known today as the *Almagest*) and by Pappus of Alexandria (in his *Mathematical Collection*). In his work, Zenodorus is said to have shown a number of results, such as

1. the area of a *regular n*-gon is greater than the area of any other *n*-gon with the same perimeter;
2. given two regular *n*-gons with the same perimeter, one with $n = n_1$, and the other with $n = n_2 > n_1$, then the regular n_2-gon has the larger area.

From these two results it is easy to see that the circle (which can be thought of as a regular "infinity-gon") with a given perimeter will have an area greater than any regular *n*-gon with the same perimeter.

We can get a mathematical "feel" for these claims with the aid of what is called the *isoperimetric quotient*. This quantity, called the I.Q., is defined for any closed curve as

$$\text{I.Q.} = \frac{A}{\pi \left(\dfrac{L}{2\pi} \right)^2} = \frac{4\pi A}{L^2}$$

where L is the perimeter of the curve and A is the area enclosed by the curve. This definition is motivated by the fact that the denominator in the first expression is the area of the circle with perimeter L, and so the I.Q. of that circle (actually, *any* circle) is 1. Thus, the isoperimetric theorem says all closed curves obey the inequality I.Q. ≤ 1 with equality iff the curve is a circle.

A similar inequality can be written in three dimensions by using the claim that, for a given surface area A, it is the sphere that has the largest volume, V. That is,

$$\frac{V}{\frac{4}{3}\pi \left(\dfrac{A}{4\pi}\right)^{3/2}} \le 1$$

with equality iff the three-dimensional body is a sphere. So, here's a pretty little problem for you to play with. First, explain what the above inequality "means," that is, where does it come from? Then, use it to derive the following interesting inequality: if $x_1, x_2, \cdots, x_n$ are n real numbers, then

$$\left(x_1^2 + x_2^2 + \cdots + x_n^2\right)^3 \ge \left(x_1^3 + x_2^3 + \cdots + x_n^3\right)^2.$$

The solution is at the end of this chapter (but don't look until you spend at least a little effort on it!).

It is instructive to calculate the numerical values of the I.Q. for some common curves, if only to see how they compare to unity. A semicircle with radius r, for example, has

$$A = \frac{1}{2}\pi r^2$$
$$L = 2r + \pi r = r(2 + \pi),$$

and so its I.Q. is

$$\frac{4\pi \left(\dfrac{1}{2}\pi r^2\right)}{r^2(2+\pi)^2} = \frac{2\pi^2}{(2+\pi)^2} = 0.7467.$$

We can generalize this a bit by computing the I.Q. of an arbitrary sector of a circle with central angle θ ($\theta = \pi$ is the special case of the semicircle). Then,

$$A = \pi r^2 \left(\frac{\theta}{2\pi}\right) = \frac{1}{2}r^2\theta$$

$$L = 2r + 2\pi r \left(\frac{\theta}{2\pi}\right) = r(2 + \theta).$$

Therefore, the I.Q. of the general circular sector is

$$\frac{4\pi \left(\frac{1}{2} r^2 \theta\right)}{r^2(2+\theta)^2} = \frac{2\pi\theta}{(2+\theta)^2}.$$

We already know the value of the I.Q. for $\theta = \pi$, but is that the largest possible value? The answer is no, and here's why.

To maximize the circular sector I.Q. is equivalent to minimizing its reciprocal, i.e., to finding that value of θ that minimizes (because we can ignore the constant "2π" factor) the expression

$$\frac{(2+\theta)^2}{\theta} = \frac{4 + 4\theta + \theta^2}{\theta} = \theta + 4 + \frac{4}{\theta}.$$

And that problem is equivalent to minimizing $\theta + 4/\theta$ because we can ignore the constant additive 4. Now, the AM-GM inequality says that

$$\theta + \frac{4}{\theta} \geq 2\sqrt{\theta \cdot \frac{4}{\theta}} = 4$$

with equality iff $\theta = 4/\theta$, i.e., iff $\theta = 2$ radians. For this θ, the I.Q. of the circular sector is

$$\frac{2\pi(2)}{(2+2)^2} = \frac{\pi}{4} = 0.7854.$$

The I.Q.'s of Zenodorus' regular n-gon's are, of course, particularly interesting, and we would expect that as $n \to \infty$, the I.Q. should approach unity (as the n-gon approaches a circle). To see that this is indeed what happens, consider figure 2.3, which shows one of the n similar triangles that a regular n-gon can be decomposed into. The two equal sides of the triangle have unity length, and the central angle is α, where

$$\alpha = 2\theta = \frac{2\pi}{n}.$$

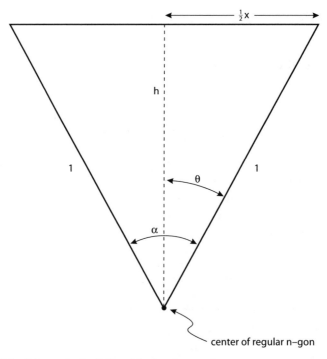

FIGURE 2.3. Triangular building block of a regular n-gon.

If we denote the height of the triangle by h and the base by x, then

$$h = \cos(\theta) = \cos\left(\frac{\pi}{n}\right)$$

$$x = 2\sin(\theta) = 2\sin\left(\frac{\pi}{n}\right).$$

The area of the triangle is A_t, where

$$A_t = \frac{1}{2}\, hx = \cos\left(\frac{\pi}{n}\right)\sin\left(\frac{\pi}{n}\right),$$

and so the area of the regular n-gon is

$$A = nA_t = n\cos\left(\frac{\pi}{n}\right)\sin\left(\frac{\pi}{n}\right).$$

The perimeter of the regular n-gon is

$$L = nx = 2n\sin\left(\frac{\pi}{n}\right),$$

and so the I.Q. of the regular n-gon is

$$I.Q._n = \frac{4\pi n \sin\left(\frac{\pi}{n}\right) \cos\left(\frac{\pi}{n}\right)}{4n^2 \sin^2\left(\frac{\pi}{n}\right)} = \frac{\pi}{n} \cot\left(\frac{\pi}{n}\right).$$

In particular, the I.Q.'s for the equilateral triangle ($n = 3$), the square ($n = 4$), the regular pentagon ($n = 5$), and the regular hexagon ($n = 6$), are:

$$I.Q._3 = \frac{\pi}{3} \cot\left(\frac{\pi}{3}\right) = 0.6046$$

$$I.Q._4 = \frac{\pi}{4} \cot\left(\frac{\pi}{4}\right) = 0.7854$$

$$I.Q._5 = \frac{\pi}{5} \cot\left(\frac{\pi}{5}\right) = 0.8648$$

$$I.Q._6 = \frac{\pi}{6} \cot\left(\frac{\pi}{6}\right) = 0.9069.$$

Thus, all regular n-gons except for the first one ($n = 3$) have I.Q.'s that exceed that of the semicircle, the I.Q. of the square is exactly equal to the I.Q. of the circular sector of maximum I.Q., and these results suggest $\lim_{n \to \infty} I.Q._n = 1$, the I.Q. of the circle. Still, these numerical results in no way prove the isoperimetric theorem. To do that, we need much deeper arguments.

As an aside, before we get into those arguments, it is interesting to note that the ancient question of how to tile the plane (how to divide an infinite two-dimensional surface into congruent n-gons) is intimately related to the concept of the I.Q. Pappus' fame today, for example, is due at least in part to his speculation that bees make their honeycombs with hexagonal cells because that structure minimizes the total wax needed to store a given amount of honey in a regular array of cells (see appendix C). This so-called "honeycomb conjecture" defied a mathematical proof until very recently (1999), when the American mathematician Thomas Hales at the University of Michigan finally succeeded in finding one. The ancients were pretty good at formulating tough problems!

A problem closely related to one that Zenodorus treated is that of inscribing the maximum area N-gon in a given circle

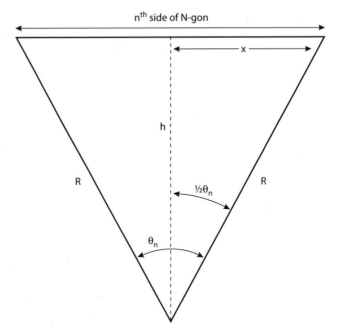

FIGURE 2.4. Making a regular N-gon.

(of radius R). The answer is that the N-gon should be regular, and the proof by modern methods is elegant. With reference to figure 2.4, θ_n is the central angle subtended by the nth side of the N-gon, which divides the N-gon into N triangles (and so $\sum_{n=1}^{N} \theta_n = 2\pi$); the nth triangle has area A_n.

To find A_n, write x (half the base of the nth triangle) as

$$x = R \sin\left(\frac{1}{2}\, \theta_n\right),$$

and h (the height of the triangle) as

$$h = R \cos\left(\frac{1}{2}\, \theta_n\right).$$

Thus,

$$A_n = xh = R^2 \sin\left(\frac{1}{2}\, \theta_n\right) \cos\left(\frac{1}{2}\, \theta_n\right),$$

or from the trigonometric identity $\sin(\alpha)\cos(\alpha) = \frac{1}{2}\sin(2\alpha)$, we have

$$A_n = \frac{1}{2}\,R^2\sin(\theta_n)$$

and so the total area of the N-gon is

$$A = \sum_{n=1}^{N}A_n = \frac{NR^2}{2}\cdot\frac{1}{N}\sum_{n=1}^{N}\sin(\theta_n)\,.$$

I'll now use a result that is a special case of a general result due to the self-taught Danish mathematician Johan L.W.V. Jensen (1859–1925), who spent his career not as an academic but rather as an engineer for the Copenhagen Telephone Company (he eventually became Chief Engineer). In 1906 he published what has come to be known as *Jensen's inequality* (you can find it stated and proven in appendix B); the special case of it that I'll use here is

$$\frac{1}{N}\sum_{n=1}^{N}\sin(\theta_n) \leq \sin\left(\frac{1}{N}\sum_{n=1}^{N}\theta_n\right)$$

with equality iff $\theta_1 = \theta_2 = \cdots = \theta_N$, i.e., when all of the central angles are equal, and so the N-gon of maximum area is the *regular* N-gon with area $(NR^2/2)\sin(2\pi/N)$.

It is immediately clear, before we get into details, that if there is a solution to the isoperimetric problem, then it must be what is called a *convex* figure. A convex figure is one that, given any two points A and B (with the requirement that these points are either on the boundary edge of the figure or inside the figure) then *all* the points on the chord $\overline{AB}$ are also either on the boundary edge or inside the figure. More graphically, the boundary edge (a convex curve) of the figure has no indentations, and there are no holes in the figure. Figure 2.5 shows two examples of nonconvex figures.

The reason why a nonconvex figure cannot be the solution to the isoperimetric problem is that it is always possible to transform such a figure into another figure (possibly still nonconvex) that has

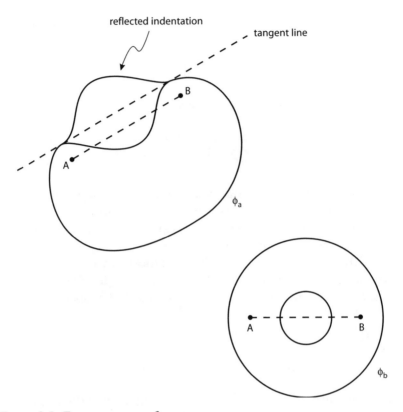

FIGURE 2.5. Two nonconvex figures.

the same (or even less) perimeter and a larger area. For example, for φ_b in figure 2.5, simply remove the hole and you have a new figure with more area and a smaller perimeter. For φ_a, reflect the indentation through the dashed tangent line (as shown), giving a new figure with increased area and the same perimeter. So, whatever the solution figure to the isoperimetric problem may be, it must be convex, and from now on we'll limit our attention to convex figures.

I'm not going to show you Zenodorus' proofs here (if you're curious you can find them in volume 2 of Heath's previously cited work *A History of Greek Mathematics*, pp. 207–12). Rather, I'll show you a more "recent" geometric analysis from the nineteenth century, due to the Swiss mathematician Jakob Steiner (1796–1863). I should tell you that in 1789 Steiner's fellow countryman Simon Lhuilier (1750–1840) also proved Zenodorus' results in a manner quite different

from the approach you'll find in Heath's book. It is Steiner's beautifully elegant 1842 arguments, however, first for the problem of Dido
and then the isoperimetric theorem, that are models of mathematical ingenuity even if they do suffer from that subtle flaw I tantalized
you about in the previous section.

Before actually presenting proofs, however, I'll conclude this section by addressing, one last time, the concept of *duality* (introduced
in the previous chapter) as it relates to the isoperimetric theorem.
The following two statements are logically equivalent, quite independent of whether or not they are actually true (they *are* true, but
that will be established in the next section):

A. Of all closed curves in a plane with equal perimeters, the
   circle bounds the largest area;
B. Of all closed curves in a plane with equal areas, the circle has
   the smallest perimeter.

To prove the claim of logical equivalency, I'll first assume that A
holds, and then show that B necessarily follows. To do this, begin
by assuming that B does *not* follow (and this will, as you'll soon
see, quickly lead to a contradiction and so B *must* follow A). Thus,
contrary to B, let's assume that for a given circle C with a given area,
there is some other closed curve D with the same area but with a
smaller perimeter.

So, imagine that we shrink C down to the smaller circle $\hat{C}$ that
has a perimeter equal to that of D. Obviously the area of $\hat{C}$ is smaller
than that of C, i.e., the area of $\hat{C}$ is smaller than the area of D. Thus,
C and D have the same perimeter but it is D, not the circle $\hat{C}$, with
the larger area, which contradicts A. This contradiction must be the
result of our using the negative of B (that B does not follow from
A), and so B *must* follow from A.

To complete this demonstration of duality we must next show the
reverse, i.e., if we assume that B holds, then A necessarily follows.
So, as before, let's assume that A does *not* follow and, as before, we'll
be able to derive a contradiction. Thus, contrary to A, let's assume
that for a given circle C with a given perimeter there is some closed
curve D with the same perimeter that has a larger area. We now
imagine that C is expanded up to the larger circle $\hat{C}$ that has an area
equal to that of D. Since we expanded C to get $\hat{C}$, then the perimeter
of C will be greater than the perimeter of C, i.e., the perimeter of C is

greater than the perimeter of D. That is, $\hat{C}$ and D have the same area but it is D, not the circle $\hat{C}$, that has the smaller perimeter, which contradicts B. This contradiction must be the result of our using the negative of A (that A does not follow from B), and so A *must* follow from B.

None of this proves the isoperimetric theorem itself, however. What we need to do next is to show either the truth of A or of B (either one will do, of course, as the other will then logically follow). That's our task in the next section.

2.3 Steiner's "Solution" to Dido's Problem

To show how Steiner arrived at his demonstration of the solution of the original problem of Dido, we need to establish two preliminary results in elementary geometry. The first one is a standard high school exercise, that of showing that any triangle inscribed in a circle, with a diameter as a side (the hypotenuse), is a right triangle. You can find a proof of this in any high school geometry text. The second result we'll need is that, of all possible triangles with two sides of given length, the triangle of maximum area is the right triangle with the given sides as the perpendicular sides. This is very easy to show. With reference to figure 2.6, let x and y be the two given sides, with angle θ between them. The height of the triangle is then $h = x\sin(\theta)$ and the area of the triangle is

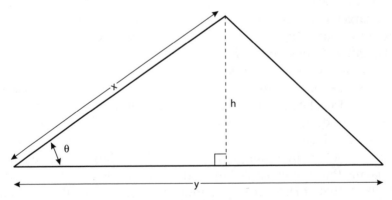

FIGURE 2.6. Maximizing the area of a triangle.

$$A = \frac{1}{2} yh = \frac{1}{2} xy \sin(\theta).$$

A is maximized, then, by making the factor $\sin(\theta)$ maximum, i.e., $\sin(\theta) = 1$, which means $\theta = 90°$, and we are done.

Now we can follow Steiner's solution to the original problem of Dido: what curve of given length joins two points on a given straight line so as to maximize the enclosed area? Let A and B be the two points on the given straight line L, as shown in figure 2.7, and suppose the solution curve C is *not* a semicircle. That means, by our first preliminary result, that there must be a point P on C such that angle $APB \neq 90°$. (Actually, for us to conclude this we should really show that the circle is the *only* curve such that for any P the angle $APB = 90°$, but I'll skip over this detail.) Figure 2.7 shows that the dashed chords AP and PB divide the area enclosed by L and C into the three regions R_1, R_2, and R_3.

Next, imagining AP and PB to be rigid rods "hinged" at P, with sliding contacts on L at A and B, let's adjust either A or B (or perhaps both) to A' and B' in such a way that the angle $A'P'B'$ is a right angle (as shown in figure 2.8). That is, as we make the adjustments $A \rightarrow A'$ and $B \rightarrow B'$, then P will move to some new point P' such that the lengths AP and $A'P'$ are equal (and also the lengths BP and $P'B'$ are equal). I'll call the resulting new curve C'; it is not necessarily a semicircle since there may be more points where the angle

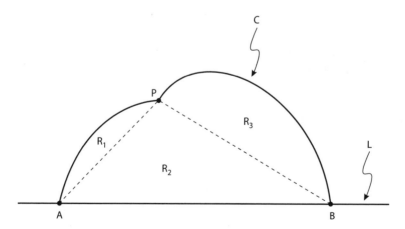

FIGURE 2.7. Steiner's isoperimetric argument, part 1.

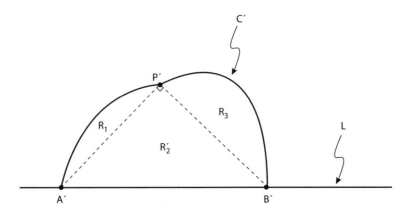

FIGURE 2.8. Steiner's isoperimetric argument, part 2.

$APB \neq 90°$). Since the two chord lengths are unchanged by these adjustments we can place regions R_1 and R_3 on the chords $A'P'$ and $P'B'$, respectively, while region R_2 will adjust to the new region R'_2.

Now, by our second preliminary result, the area of R'_2 is greater than the area of R_2. So, we have taken an arbitrary curve C and transformed it into a curve C' *with the same perimeter* that encloses (with L) an area greater than that enclosed by C and L. (We do have another potential objection here; how do we know that the R_1 and R_2 regions don't "bump into" each other, i.e., overlap, during the adjustment? We don't, but again I'll pass over this detail.) The only curve C that does not allow such an area-increasing, perimeter-preserving transformation is the semicircle (as then there is no point P on C such that the angle $APB \neq 90°$).

At this point Steiner believed he had constructed a purely geometric solution to the original problem of Dido, and he proceeded to attack the isoperimetric problem by next making the following preliminary observation: if φ is the (convex) solution figure to the isoperimetric problem, then any chord joining two points on the boundary of φ that bisects the perimeter (assuming such a chord exists) would also bisect the area. (The existence of such a perimeter-bisecting chord is demonstrated in appendix D.) This is so because if we suppose the area is not bisected, then we could take the larger area and reflect it through the chord. That would give us a new figure with the same perimeter and a larger area than that of φ, which is impossible because it is φ (by assumption) that is the solution figure.

Now, given the solution figure φ to the isoperimetric problem, we draw a chord that bisects the perimeter of φ, as well as, as shown above, the area of φ. This bisection splits φ into φ_1 and φ_2, where φ_1 and φ_2 have equal areas *and* equal perimeters (indeed, they have a shared perimeter, since the bisection chord is common to φ_1 and φ_2). We clearly maximize the area of φ by maximizing the (equal) areas of φ_1 and φ_2. Notice, however, that φ_1 and φ_2 are each semicircular disks by Steiner's original solution to the Dido question, and so φ is a circular disk. So, concluded Steiner, the solution curve to the isoperimetric problem is a circle.

2.4   How Steiner Stumbled

Steiner's analysis is undeniably clever. His contemporary, however, the German mathematician Peter Dirichlet (1805–59), pointed out that in addition to the objections I mentioned in the previous section, Steiner had made the unstated assumption that there actually is, in fact, a solution to the isoperimetric problem. Most people reply to that objection by saying "*Of course* there is a solution—it's *obvious* there is a solution!" That was Steiner's reply, in fact, at least at first, but it ignores the fact that there are plenty of geometry questions that look at first glance like they should have solutions—but in fact do not.

Consider, for example, the problem of finding that convex figure of greatest area among all convex figures with a perimeter less than 1. The fact is that there is no solution to this problem; here's why. Let ε be some arbitrarily small positive number, and suppose that the convex figure φ has perimeter $1 - \varepsilon$. Then, simply expand φ up in scale to a new (similar) figure with the larger perimeter $1 - \frac{1}{2}\varepsilon$ (which of course is still less than 1). This new figure has an area greater than that of φ, and in fact we can repeat this process as many times as we wish. That is, we can generate an endless sequence of convex figures all with perimeters less than 1 but with ever increasing areas; there is no "largest area" figure, and so we would be in error to a priori assume that there is a solution figure.

An even more dramatic illustration of the danger in assuming the existence of a solution is the surprise answer to a problem that is of a nature opposite to the one just considered. Called the "Kakeya problem" after S. Kakeya (1886–1947), the Japanese mathematician who posed it in 1917, it asks: what is the smallest area in which a line segment of unit length (in arbitrary units) can be rotated through 360°? Virtually everybody believes, upon first hearing this, that there is a smallest area. Kakeya himself conjectured that the minimum area is $\frac{1}{8}\pi$. In a paper published in 1928, however, the Russian mathematician Abram Besicovitch (1891–1970) showed that no matter how long the line segment, there is *no* smallest area! That is, there *does* exist a figure with the area (for example) of the period at the end of this sentence in which a line segment one million light years long can be rotated through 360°. Besicovitch actually showed how to make such a figure (it's nonconvex and *very* complicated— no surprise there!), and you can find an elementary discussion of just how to construct it in Besicovitch's own words, in his paper "The Kakeya Problem" (*American Mathematical Monthly*, September 1963, pp. 697–706). Besicovitch's result shows that one must be very careful before assuming there is always a solution.

I'll close this section with the observation that the entire point of one of the great, historically important maximum problems in pure number theory was to prove that there is *no* maximum. This is Euclid's wonderful demonstration, in his *Elements* (Book 9), that there is no largest prime, i.e., that there is an infinity of integers with no factors other than themselves and unity. His elegant proof is perhaps the ultimate in simplicity. Suppose that there *are* only n primes, labeled $p_1, p_2, \cdots, p_n$. That is, p_n is the largest prime. Then, form the new (obviously much larger) integer:

$$P = p_1 p_2 \cdots p_n + 1.$$

What can we say about P?

By our assumption that p_n is the largest prime, we conclude that P must *not* be prime. It therefore must be possible to write P as the product of primes (simply keep factoring the factors of P until all the factors are prime), but clearly *none* of the assumed finite number of primes divides P (because of that "+1"). So, P is *not* factorable into a product of primes, which says P itself must be a prime. But that contradicts our assumption that $p_n(<P)$ is the largest prime. The only way out of this swamp is to admit that our assumption is false and that there is no largest prime; there is an infinity of primes.

Before leaving the primes, let me show you just one more "there is no maximum" fact about primes that surprises most people when they first encounter it. Since the primes are infinite in number, then of course no matter how far up we go in the integers we will always keep finding them. But that doesn't mean they occur in any sort of regular way. Far from it! Indeed, if we call $g(p_n)$ the *gap* between consecutive primes p_n and p_{n+1}, i.e., if we write $g(p_n) = p_{n+1} - p_n - 1$, then in fact $g(p_n)$ has no maximum. There are *always* two consecutive primes such that the gap between them is as large as we like. For example, if g is to be at least 10^{100} (the famously huge googol), or if it is to be the even more impressive 10^{googol} (the equally famous but stupendously larger googolplex) then there exist two consecutive primes that have a gap as least as large as those g's.

The proof is direct: the production of a *specific* sequence of consecutive integers with a length g that is obviously free of primes (every integer in the sequence is evenly divisible.) Simply take the desired value of g and form the sequence of consecutive integers of length g defined by

$$(g+1)! + 2, \quad (g+1)! + 3, \quad (g+1)! + 4, \cdots, \quad (g+1)! + g + 1.$$

The first integer is divisible by 2, the second is divisible by 3, etc., etc., etc. and the last integer is divisible by $g+1$. Since we could have started with g as *any* finite integer, then there is no maximum g.

Even though there are arbitrarily large gaps between successive primes, it has also been shown that successive primes do obey a certain rule on when they *must* occur. In 1845 the French mathematician Joseph Bertrand (1822–1900) conjectured that for all $n > 3$ there is at least one prime between n and $2n - 2$ (the conjecture is often stated in the alternative— and perhaps more elegant—form of for all $n > 1$ there is at least one prime between n and $2n$). Bertrand's conjecture was proven in 1850, by the Russian mathematician Pafnuty Chebyshev (1821–94). Arbitrarily large gaps are compatible with this result because large gaps occur only when the numbers in the gap are *vastly* greater than the gap length.

In 1932 a much simpler proof of Bertrand's conjecture was found by the Hungarian mathematician Paul Erdõs (1913–96), when he was but an eighteen-year-old student in Budapest. When Erdõs announced his proof, he accompanied it with the rhyme "Chebyshev said it, and I say it again, there is always a prime between n and $2n$."

2.5   A "Hard" Problem with an Easy Solution

Here's an elegant solution to an interesting problem that occurs in many *advanced* books on calculus, which we can attack using the elementary concepts developed earlier in this chapter. Suppose we are presented with a length of string, which we are to cut into two pieces. Then, with those two pieces we are to form two figures with *prescribed shapes*, e.g., a square and a circle, or a half-circle and an equilateral triangle. How should we cut the string to minimize the total area of the two figures? Or, what if instead of just one cut and two figures we are more generally to cut the string $n - 1$ times and then to form n figures (with prescribed shapes) enclosing minimum total area? The general question sounds like a tough problem (our first question is fairly easy) but, perhaps astonishingly, we can solve the general case easily, too, with the aid of the isoperimetric quotient (I.Q.) and Jensen's inequality (see appendix B).

First, recall from the first part of this chapter that every planar figure *shape*, independent of its actual size, has the *same* I.Q., defined as

$$I.Q. = \frac{4\pi A}{L^2},$$

where A is the area of the figure and L is the figure's perimeter. If we write $1/\lambda_i$ as the I.Q. of the ith prescribed figure (and so λ_i is a given), and if A_i and L_i denote the area and the perimeter, respectively, of that figure, then the total enclosed area is A, where

$$A = A_1 + A_2 + \cdots + A_n = \frac{1}{4\pi} \left(\frac{L_1^2}{\lambda_1} + \frac{L_2^2}{\lambda_2} + \cdots + \frac{L_n^2}{\lambda_n} \right).$$

If we write L as the total uncut length of the string, then of course

$$L_1 + L_2 + \cdots + L_n = L, \qquad \text{with all } L_i > 0,$$

and it is with this *constraint* that we wish to find the L_i that minimize A. This is the sort of problem usually treated with a calculus technique called *Lagrange multipliers* (discussed in chapter 6), but we can do it now with no calculus, using Jensen's inequality. (The calculus approach *is* much faster, so don't conclude that calculus isn't important!)

Applying Jensen's inequality to the strictly convex function $f(x) = x^2$, we have (with all $c_i > 0$ and summing to one)

$$(c_1 x_1 + c_2 x_2 + \cdots + c_n x_n)^2 \le c_1 x_1^2 + c_2 x_2^2 + \cdots + c_n x_n^2,$$

with equality iff $x_1 = x_2 = \cdots = x_n$. So, define c_i and x_i as

$$c_i = \frac{\lambda_i}{\lambda_1 + \lambda_2 + \cdots + \lambda_n},$$

$$x_i = \frac{L_i}{\lambda_i}.$$

Note that $c_i > 0$ for any i and that, indeed, the c_i sum to 1.

Our inequality then becomes

$$\left[\frac{\lambda_1}{\lambda_1 + \lambda_2 + \cdots + \lambda_n} \cdot \frac{L_1}{\lambda_1} + \frac{\lambda_2}{\lambda_1 + \lambda_2 + \cdots + \lambda_n} \cdot \frac{L_2}{\lambda_2} \right. $$
$$\left. + \cdots + \frac{\lambda_n}{\lambda_1 + \lambda_2 + \cdots + \lambda_n} \cdot \frac{L_n}{\lambda_n} \right]^2$$

$$\le \frac{\lambda_1}{\lambda_1 + \lambda_2 + \cdots + \lambda_n} \left(\frac{L_1}{\lambda_1} \right)^2 + \frac{\lambda_2}{\lambda_1 + \lambda_2 + \cdots + \lambda_n} \left(\frac{L_2}{\lambda_2} \right)^2$$
$$+ \cdots + \frac{\lambda_n}{\lambda_1 + \lambda_2 + \cdots + \lambda_n} \left(\frac{L_n}{\lambda_n} \right)^2,$$

or, after some canceling,

$$\frac{(L_1 + L_2 + \cdots + L_n)^2}{(\lambda_1 + \lambda_2 + \cdots + \lambda_n)^2} \leq \frac{1}{(\lambda_1 + \lambda_2 + \cdots + \lambda_n)} \left[\frac{L_1^2}{\lambda_1} + \frac{L_2^2}{\lambda_2} + \cdots + \frac{L_n^2}{\lambda_n} \right],$$

or, again after some canceling,

$$\frac{(L_1 + L_2 + \cdots + L_n)^2}{\lambda_1 + \lambda_2 + \cdots + \lambda_n} = \frac{L^2}{\lambda_1 + \lambda_2 + \cdots + \lambda_n}$$

$$\leq \frac{L_1^2}{\lambda_1} + \frac{L_2^2}{\lambda_2} + \cdots + \frac{L_n^2}{\lambda_n} = 4\pi A,$$

with equality iff $x_1 = x_2 = \cdots = x_n$.

So, A is minimized (becomes *equal* to its lower bound) when

$$\frac{L_1}{\lambda_1} = \frac{L_2}{\lambda_2} = \cdots = \frac{L_n}{\lambda_n} = \alpha, \qquad \text{to be determined next.}$$

That is, when A is minimized, we have A equal to

$$\frac{L^2}{4\pi (\lambda_1 + \lambda_2 + \cdots + \lambda_n)} = \frac{1}{4\pi} \left(\frac{L_1^2}{\lambda_1} + \frac{L_2^2}{\lambda_2} + \cdots + \frac{L_n^2}{\lambda_n} \right),$$

or

$$\frac{L^2}{4\pi (\lambda_1 + \lambda_2 + \cdots + \lambda_n)} = \frac{1}{4\pi} \alpha (L_1 + L_2 + \cdots + L_n) = \frac{\alpha L}{4\pi},$$

and so

$$\alpha = \frac{L}{\lambda_1 + \lambda_2 + \cdots + \lambda_n}.$$

Thus,

$$\frac{L_1}{\lambda_1} = \frac{L_2}{\lambda_2} = \cdots = \frac{L_n}{\lambda_n} = \frac{L}{\lambda_1 + \lambda_2 + \cdots + \lambda_n},$$

or, at last,

$$L_i = \frac{\lambda_i}{\lambda_1 + \lambda_2 + \cdots + \lambda_n} L, \qquad i = 1, 2, 3, \cdots, n,$$

which is the solution to the problem of how to cut the original string of length L into n pieces, each of length L_i. For example, suppose we are to cut the string into two pieces and use those pieces to form the circle and the half-circle that enclose minimum total area. As shown earlier, λ_1 (circle) $= 1/1 = 1$, and λ_2 (semicircle) $= 1/0.7467$. Thus,

$$L_1 = \frac{1}{1 + \dfrac{1}{0.7467}} \, L = \frac{0.7467}{1.7467} \, L = 0.4275 \, L,$$

and so, to minimize the total area of the two figures, 42.75% of the string should go to the circle and the rest should go to the semicircle.

2.6 Fagnano's Problem

To end this chapter I'll describe two fascinating examples of geometric minimization; in the first the demonstration of the existence of a solution is explicit. To begin, consider the so-called "minimum-perimeter triangle of Fagnano." This problem has its origins with the Italian mathematician Giulio Carlo Toschi di Fagnano (1682–1766), who showed the existence part, and his priest-mathematician son Giovanni Francesco Fagnano (1715–97), who completed the minimization argument in 1775. The father's contribution was to show, given any acute-angled triangle ABC (as shown in figure 2.9) and any *given* point U on one of the sides (BC in the figure), how to construct the inscribed triangle of minimum perimeter *with a vertex*

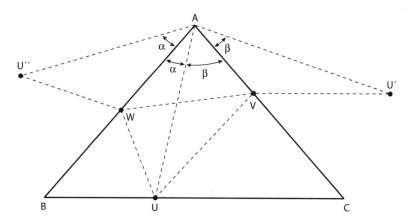

FIGURE 2.9. Fagnano's problem, part 1.

at U. (You'll soon see why the restriction of an *acute* triangle is necessary). The son later showed how to pick *U* to select the *absolute* minimum-perimeter triangle. For this problem, then, there is no question about the existence of a solution. The son used the differential calculus to arrive at his answer, but the clever geometric proof that follows is due to the Hungarian mathematician Lipót Fejér (1880–1959), who discovered it while still a student at the University of Budapest.

To see the existence of a solution for a given *U*, first connect *U* to vertex *A* to form line segment *AU*, and then "reflect" *AU* about the triangle sides *AB* and *AC* to form the line segments *AU″* and *AU′*, respectively. Then, with *W* and *V* as arbitrary points on *AB* and *BC*, respectively, connect *U″* to *W* and *U′* to *V*. And finally, connect *W*, *V*, and *U* to form an inscribed triangle. By this construction it is clear that (in terms of length) $WU'' = WU$, and also that $U'V = UV$. Now, the perimeter of the inscribed triangle *UVW* is $UV + VW + WU$, but this is equal to $U'V + VW + WU''$, the length of the broken line connecting *U′* to *V* to *W* to *U″*. The length of the *broken* line will be minimized when, instead of being broken, it is straight. That is, after reflecting *AU* about the sides *AB* and *AC*, we can determine the *W* and the *V* that minimize the perimeter of the inscribed triangle *UVW* by simply connecting *U″* and *U′* with a straight line and observing where that line intersects *AB* and *AC*, respectively, as shown in figure 2.10. Thus, we have found by construction the unique inscribed minimum-perimeter triangle *UVW for a given U*.

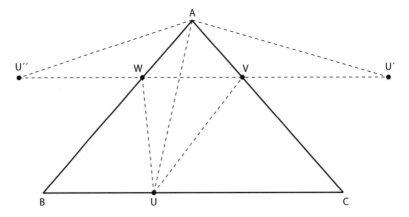

FIGURE 2.10. Fagnano's problem, part 2.

The final part of the problem is to determine that particular U that gives the minimum of the minimums. Notice, first, that by construction the triangle $AU'U''$ is an isosceles triangle, with the angle at vertex A equal to $2\alpha + 2\beta = 2(\alpha + \beta)$, i.e., this angle is always twice the vertex A angle of the original ABC triangle, and so the vertex A angle of $AU'U''$ is the same for *any* choice of the point U. Thus, the "best" choice for U, the particular U that minimizes the perimeter of UVW, is the U that minimizes the equal length sides AU'' and AU' of the isosceles triangle $AU'U''$. That is so because, given an isosceles triangle with a fixed vertex angle at A, we minimize the base of that triangle ($U''WVU'$, equal to the inscribed triangle's perimeter) by minimizing the lengths of the two equal sides of the isosceles triangle. But that simply says we pick U to minimize the length of AU, i.e., we should draw AU perpendicular to BC. In other words, when we have the best U, we have AU as the *altitude* from vertex A to the side BC.

If you look back at what we have done you'll see that the points W and V are uniquely determined, i.e., the minimum-perimeter inscribed triangle is unique. The immediate consequence of this is that we don't have to go through all of the detailed steps of the proof (e.g., reflecting lines about other lines) to actually draw the minimum-perimeter triangle. This is because our choice of the side BC to work from was arbitrary—we could have started with side AC and found that the resulting line BU would be the altitude from vertex B to AC. Or we could have started with side AB and found that the resulting line CU would be the altitude from vertex C to AB. In the end, however, we would arrive at the *same* inscribed minimum-perimeter triangle because that triangle is unique. So, to actually construct UVW, simply draw the three altitudes and thereby immediately locate the points U, V, and W. The resulting inscribed triangle is called the *pedal* or *orthic* triangle of the original triangle ABC. You can also now see why ABC must be acute—it insures that all three altitudes are *inside* ABC, i.e., that U, V, and W lie on the sides of ABC and so the triangle UVW is truly an *inscribed* triangle.

As my final example to demonstrate that interest in geometric minimization did not cease with the ancients, consider the problem of the "spanning circle of n points." Imagine that you have n points positioned arbitrarily in the plane. We can measure the distance between every possible pair of points and call the maximum distance d. There are, of course, just $\frac{1}{2}n(n-1)$ such distances, and so computing the value of d is a straightforward matter. Now, suppose we wish to draw a circle whose interior contains all n points; such a circle is said to *span* the points. The problem is to determine the smallest circle that spans the points. A practical form of this problem would be, for example, determining where to locate a fire station within a community to minimize the maximum distance from the fire station to any of the surrounding homes. It is clear, of course, that a circle with radius d spans the points; simply pick *any one* of the points as the center of a circle with radius d and observe that, by definition, no other point is more distant than d.

Is it possible to construct a spanning circle that is smaller? Yes, indeed it is. A spanning circle with a radius no greater than $\frac{1}{3}\sqrt{3}d \approx$ 0.577 d always exists. The proof is by elementary (but ingenious) geometry, and you can find it all worked out in the book by Hans Rademacher and Otto Toeplitz, *The Enjoyment of Mathematics* (Princeton University Press 1957, pp. 103–10). Even more on the minimum spanning circle—which dates back to at least 1860 and the work of the English mathematician J. J. Sylvester (1814–97)—can be found in the book by Franco P. Preparata and Michael Ian Shamos, *Computational Geometry* (Springer-Verlag 1985, pp. 248–54). Those pages also discuss the dual problem: what is the *largest* empty circle inside the convex hull of the given n points (think of the points as vertical posts, and a rubber band snapped all around them, as shown in

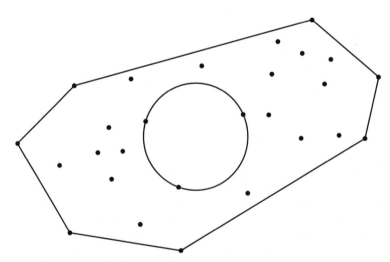

FIGURE 2.11. A convex hull and its largest interior empty set.

figure 2.11) that contains none of the points? That would tell us, for example, where to place an *objectional* service facility for the town, e.g., a centrally located waste-treatment plant that nobody wants to live near!

Solution to the Problem in Section 2.2

A sphere of radius r has surface area $A = 4\pi r^2$. Thus,

$$r = \left(\frac{A}{4\pi} \right)^{1/2},$$

and so its volume is

$$V = \frac{4}{3}\pi r^3 = \frac{4}{3}\pi \left(\frac{A}{4\pi} \right)^{3/2}.$$

The claim is that, for a given A, *this V* (for a sphere) is the largest possible. So, if V is the volume of *any* three-dimensional body, then

$$\frac{V}{\frac{4}{3}\pi \left(\frac{A}{4\pi} \right)^{3/2}} \leq 1,$$

with only the sphere achieving equality (see the end of this box).

For the second part of this problem, suppose we have n spheres with radii $x_1, x_2, \cdots, x_n$. The total surface area and total enclosed volume are

$$A = 4\pi x_1^2 + 4\pi x_2^2 + \cdots + 4\pi x_n^2 = \sum_i 4\pi x_i^2$$

$$V = \frac{4}{3}\pi x_1^3 + \frac{4}{3}\pi x_2^3 + \cdots + \frac{4}{3}\pi x_n^3 = \sum_i \frac{4}{3}\pi x_i^3.$$

Now, imagine that we glue all of these spheres together to form a (rather lumpy!) single body. This single body will obey the above inequality (which, after squaring and rearranging, becomes $A^3 \geq 36\pi V^2$). So,

(continued)

$$\left(\sum_i 4\pi x_i^2\right)^3 \geq 36\pi \left(\sum_i \frac{4}{3}\pi x_i^3\right)^2,$$

which quickly reduces to

$$\left(\sum_i x_i^2\right)^3 \geq \left(\sum_i x_i^3\right)^2.$$

Notice that this argument only makes physical sense if all of the $x_i \geq 0$, because a physical sphere can't have a negative radius. However, if one or more of the $x_i < 0$, it is clear that the left-hand side of the inequality is indifferent to the sign, while the right-hand side becomes *smaller*. That is, one or more $x_i < 0$ simply strengthens the inequality. Thus,

$$\left(x_1^2 + x_2^2 + \cdots + x_n^2\right)^3 \geq \left(x_1^3 + x_2^3 + \cdots + x_n^3\right)^2$$

for *all* real x_i.

Historical note: The entire argument of this box is based on the assumed truth of the three-dimensional isoperimetric theorem, i.e., on the inequality $A^3 \geq 36\pi V^2$. This was formally established in 1884 by the German mathematician H. A. Schwarz (1843–1921). The general n-dimensional isoperimetric inequality was later shown to be

$$A^n \geq \frac{2\pi^{\frac{n}{2}} n^{n-1} V^{n-1}}{\Gamma\left(\frac{n}{2}\right)},$$

where Γ is Euler's generalization (with his gamma function integral) of the factorial function:

$$\Gamma(x) = \int_0^\infty e^{-t} t^{x-1} dt.$$

The general isoperimetric inequality was established in 1939 by the German mathematician Erhard Schmidt (1876–1959).

3.

Medieval Maximization
and Some Modern Twists

3.1 The Regiomontanus Problem

After the ancient isoperimetric problems discussed in the previous chapter, it seems that very little if anything new on minimiza-
tion/maximization theory appeared in mathematics for a very long
time. Indeed, not for another *fifteen centuries* after Christ! And then,
in 1471, the German mathematician Johann Müller (1436–76), more
commonly known today as Regiomontanus, posed a clever maximization problem totally unlike any that had come before. I'll state
it here in slightly more dramatic fashion than he did, but the basic
problem itself is as Regiomontanus conceived it.

A somewhat confusing trait of some of the medieval mathematicians makes it appear that there were more of them than
there were—they often used more than one name. In the case
of Johann Müller, for example, who was born in Königsberg
(which mean "King's Mountain"), he Latinized that to "Re-
gio monte," which soon evolved into Regiomontanus. Two
other famous examples of the double-named syndrome are
the Italians Leonardo of Pisa (1170–circa 1250), also known
as "Fibonacci," and Niccolo Fontana (1500–57), who was also
called "Tartaglia." So, we have six names, but only three mathematicians.

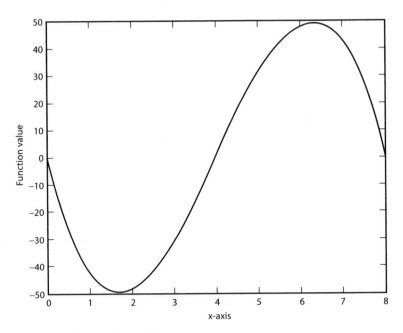

FIGURE 3.1. Tartaglia's cubic function.

As an aside on Tartaglia, a moderately interesting maximization problem is due to him, dating from some time between 1556 and 1560; but it is, at heart, really only a slightly more sophisticated version of Euclid's ancient problem of dividing a number into two parts to maximize their product. The Regiomontanus problem is far more advanced, for two reasons: it is motivated by a *physical* setting, and it requires the use of a trigonometric function. In Tartaglia's abstract, algebraic problem, we are to divide 8 into two parts so that the product of their product and their difference is maximized. Thus, if the parts are x and $8 - x$, then we are to maximize $x(8 - x)[x - (8 - x)] = -2x^3 + 24x^2 - 64x$, where of course $0 \leq x \leq 8$. Tartaglia almost certainly structured the statement of this problem with the intent of arriving at a cubic; see my book *An Imaginary Tale: The Story of $\sqrt{-1}$* (Princeton University Press 1998), for Tartaglia's part in the history of the cubic equation. He did not reveal his method of solution, but he did publish the correct answer: $x = 4(1 + 1/\sqrt{3}) = 6.309401$ (see figure 3.1). For how he *might* have reasoned, see V. M. Tikhomirov, *Stories about Maxima and Minima* [translated from the Russian] (The American Mathematical Society 1990, pp. 37–39).

A painting is hung flat against an art museum wall, with its bottom and top edges at distances a and b, respectively, from the floor. That is, the vertical dimension of the painting is $(b - a)$. The painting is viewed by a tourist whose eye level is distance h from the floor, where $h < a$. That is, the picture is hung high on the wall to avoid the front of a crowd of tourists from blocking the view of those in the back. How far from the wall should a tourist stand to maximize the viewing angle subtended at his eye by the painting, i.e., so that the painting appears as large as possible? Figure 3.2 shows the geometry of the problem, and introduces our notation. (Note that the figure shows that the condition $h > a$ leads immediately— by inspection—to the "uninteresting" result that, to maximize his viewing angle, the tourist should stand at $x = 0$, i.e., with his nose pushed hard into the painting! The geometry says the "viewing angle" is 180°, but it seems clear he would not enjoy the view.)

Mathematically, our problem is simply that of determining the value of x that maximizes the angle $\theta = \alpha - \beta$. Today this problem is popular with the authors of calculus textbooks, but the year

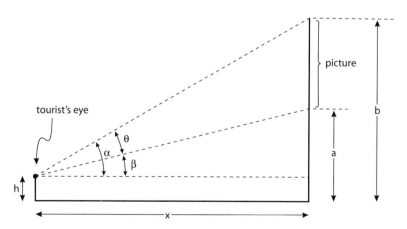

FIGURE 3.2. Regiomontanus' hanging picture (one-dimensional).

1471 was still a couple of hundred years short of the beginnings of
the differential calculus. That's why the original solution (it is not
known if it is due to Regiomontanus himself) was in the form of a
complicated (in my opinion) geometric construction; you can find
it discussed in the book by Ivan Niven, *Maxima and Minima without
Calculus* (The Mathematical Association of America 1981, pp. 71–
72). What I'll show you here, instead, is a very clever noncalculus
solution that is only slightly more general than the one given in
Eli Maor's *Trigonometric Delights* (Princeton University Press 1998,
pp. 46–48). Later, in section 3.5, I'll make the solution a bit more
realistic (and complicated, which will require a computer to give us
numerical results).

The essential idea is to maximize $\tan(\theta) = \tan(\alpha - \beta)$, which
is equivalent to our problem of maximizing $\theta = \alpha - \beta$ since the
tangent function monotonically increases as its argument increases.
We begin, then, with the trigonometric identity

$$\tan(\theta) = \tan(\alpha - \beta) = \frac{\tan(\alpha) - \tan(\beta)}{1 + \tan(\alpha)\tan(\beta)},$$

where, from figure 3.2, we have

$$\tan(\alpha) = \frac{b - h}{x}$$

$$\tan(\beta) = \frac{a - h}{x}.$$

So,

$$\tan(\theta) = \frac{\dfrac{b - h}{x} - \dfrac{a - h}{x}}{1 + \dfrac{b - h}{x} \cdot \dfrac{a - h}{x}} = \frac{(b - h)x - (a - h)x}{x^2 + (b - h)(a - h)}$$

$$= \frac{(b - a)x}{x^2 + (b - h)(a - h)}.$$

To maximize $\tan(\theta)$, and hence θ itself, I'll use the trick of minimizing its reciprocal, i.e., let's examine the function

$$\frac{1}{\tan(\theta)} = \frac{x^2 + (b - h)(a - h)}{(b - a)x} = \frac{x}{b - a} + \frac{(a - h)(b - h)}{x(b - a)}$$

and ask for what value of x do we have a minimum?

To answer that question, recall the AM-GM inequality. For any two positive numbers, y_1 and y_2, the AM-GM inequality says $y_1 + y_2 \geq 2\sqrt{y_1 y_2}$, with equality iff $y_1 = y_2$. Thus, setting

$$y_1 = \frac{x}{b-a}$$

$$y_2 = \frac{(a-h)(b-h)}{x(b-a)},$$

we have

$$\frac{1}{\tan(\theta)} \geq 2\sqrt{\left[\frac{x}{b-a}\right]\left[\frac{(a-h)(b-h)}{x(b-a)}\right]} = 2\sqrt{\frac{(a-h)(b-h)}{(b-a)^2}},$$

with equality iff

$$\frac{x}{b-a} = \frac{(a-h)(b-h)}{x(b-a)}.$$

That is, $1/\tan(\theta)$ is never less than the *constant* $[2/(b-a)]$ $\sqrt{(a-h)(b-h)}$ and is equal to that constant iff $x = \sqrt{(a-h)(b-h)}$. This value of x minimizes $1/\tan(\theta)$ or, to say the equivalent, *maximizes* $\tan(\theta)$, which means θ itself is maximized. For this value of x, the value of $\tan(\theta)$ is (using the expression for $\tan(\theta)$ given in the previous paragraph)

$$\frac{(b-a)\sqrt{(a-h)(b-h)}}{\left(\sqrt{(a-h)(b-h)}\right)^2 + (b-h)(a-h)} = \frac{(b-a)\sqrt{(a-h)(b-h)}}{2(a-h)(b-h)}$$

$$= \frac{b-a}{2\sqrt{(a-h)(b-h)}}.$$

So, the answer to the Regiomontanus problem is that the tourist should stand away from the wall by the distance

$$x = \sqrt{(a-h)(b-h)}$$

and, at that distance, he will experience the maximum viewing angle of

$$\theta_{max} = \tan^{-1}\left\{\frac{b-a}{2\sqrt{(a-h)(b-h)}}\right\}.$$

A special, amusing case of interest is that of the "bug's-eye view," with $h = 0$. Then,

$$x = \sqrt{ab}$$

$$\theta_{max} = \tan^{-1}\left\{\frac{b-a}{2\sqrt{ab}}\right\}.$$

For example, if we have a large painting hung such that $a = 8$ feet and $b = 20$ feet, then a bug on the floor should position itself at a distance of

$$\sqrt{8 \cdot 20}\ \text{ft} = 12.65\ \text{ft}$$

and, at that distance from the wall, it will enjoy a viewing angle of

$$\theta_{max} = \tan^{-1}\left\{\frac{20-8}{2\sqrt{160}}\right\} = 25.4°.$$

As a less whimsical example, suppose a six-foot-tall adult comes to the museum with his three-foot-tall child. To maximize their individual views of that same painting, each will of course stand at a different distance and, perhaps even more interestingly, each will experience a significantly different maximized viewing angle. So, the optimal viewing distances for each are

$$\text{adult:}\quad \sqrt{(8-6)(20-6)}\ \text{ft} = 5.29\ \text{ft}$$

$$\text{child:}\quad \sqrt{(8-3)(20-3)}\ \text{ft} = 9.22\ \text{ft}$$

while the maximized viewing angles for each are

$$\text{adult:}\quad \theta_{max} = \tan^{-1}\left\{\frac{20-8}{2\sqrt{28}}\right\} = 48.6°$$

$$\text{child:}\quad \theta_{max} = \tan^{-1}\left\{\frac{20-8}{2\sqrt{85}}\right\} = 33.1°.$$

The adult sees a nearly 50% larger painting (in the vertical direction) than does the child.

An amusing little twist on these calculations appeared in the May 1984 issue of the *American Journal of Physics*. There, as a challenge problem for readers, the question called for the calculation of the distance a man (wearing trousers of length ℓ) should stand away from a dressing room mirror to have the best view of his trousers, if his eyes are distance h above the floor. There was no historical discussion given, but you can now see that it is just a slight variation on the original Regiomontanus problem. It was solved in the *AJP* using calculus, but it is easily handled with the AM-GM inequality, just as in the last analysis. See if you can do it (the answer is at the end of this chapter).

3.2 The Saturn Problem

A very interesting, somewhat more complicated, variation on the original Regiomontanus problem is the not so well known Saturn problem. It doesn't yield to the AM-GM inequality, but we will still be able to find a pretty solution. For this new problem, the viewer is imagined to be on the surface of a (spherical) planet that has a ring—which, for the solar system, of course means Saturn. If we further imagine that we measure latitude upward from the plane that contains the ring (see figure 3.3), then the latitude α increases from 0° in the ring plane up to 90° at the geographical north pole.

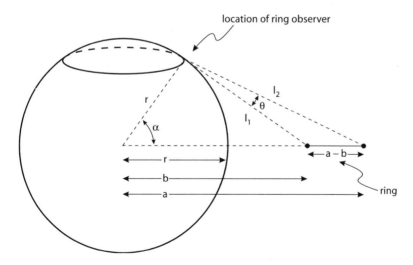

FIGURE 3.3. Geometry of the Saturn problem.

If we denote the radius of the planet by r, and the outer and inner radii of the ring by a and b, respectively, then our problem is to find that value of α at which the observed angular width θ of the ring is maximum. This value, of course, actually determines two circles of latitude all around the planet that give the maximum viewing angle, one above the ring plane (as shown in figure 3.3) and one symmetrically positioned below the ring plane.

The geometry of the Saturn problem is really quite straightforward. In the notation of figure 3.3 we have, from a triple application of the law of cosines,

$$(a - b)^2 = \ell_1^2 + \ell_2^2 - 2\ell_1\ell_2 \cos(\theta)$$
$$\ell_2^2 = r^2 + a^2 - 2ra \cos(\alpha)$$
$$\ell_1^2 = r^2 + b^2 - 2rb \cos(\alpha).$$

So,

$$
\begin{aligned}
\cos(\theta) &= \frac{\ell_1^2 + \ell_2^2 - (a - b)^2}{2\ell_1\ell_2} \\[2mm]
&= \frac{r^2 + b^2 - 2rb\cos(\alpha) + r^2 + a^2 - 2ra\cos(\alpha) - (a - b)^2}{2\ell_1\ell_2} \\[2mm]
&= \frac{2r^2 + a^2 + b^2 - 2r\cos(\alpha)(a + b) - (a^2 - 2ab + b^2)}{2\ell_1\ell_2} \\[2mm]
&= \frac{2r^2 - 2r\cos(\alpha)(a + b) + 2ab}{2\ell_1\ell_2} = \frac{r^2 + ab - r(a + b)\cos(\alpha)}{\ell_1\ell_2}.
\end{aligned}
$$

Since

$$\ell_1\ell_2 = \sqrt{\{r^2 + a^2 - 2ra\cos(\alpha)\}\{r^2 + b^2 - 2rb\cos(\alpha)\}},$$

then we have

$$\theta = \cos^{-1}\left[\frac{r^2 + ab - r(a + b)\cos(\alpha)}{\sqrt{\{r^2 + a^2 - 2ra\cos(\alpha)\}\{r^2 + b^2 - 2rb\cos(\alpha)\}}}\right].$$

This expression for the angle subtended by the ring at the observer's eye makes sense, of course, only as long as α is such that the entire ring is visible. Too large an α would require "looking through the ground" to see the inner edge of the ring. We can calculate the maximum value of α at which the inner ring edge is still visible by observing that at that α (call it $\hat{\alpha}$), the line-of-sight from the surface of the planet to the inner ring edge is tangent to the surface of the planet. That is, the radius to the location of the observer is perpendicular to ℓ_1, and so

$$\cos(\hat{\alpha}) = \frac{r}{b},$$

or

$$\hat{\alpha} = \cos^{-1}\left(\frac{r}{b}\right).$$

The case of Saturn (with the values $r = 56{,}900$ km, $a = 138{,}800$ km, and $b = 88{,}500$ km), gives us

$$\hat{\alpha} = \cos^{-1}\left(\frac{56{,}900}{88{,}500}\right) = 49.99°,$$

and so we need consider only the values of θ that occur for the interval $0° \leq \alpha \leq 49.99°$.

And by "consider" I mean that this formulation of the problem literally demands a computer analysis. That is, let's simply plot θ as we let α vary from $0°$ to $49.99°$. If there is a maximum for θ in this interval (where the entire ring is visible) we'll see it in the plot. This is an approach not easily available to precomputer-age analysts, of course, but today it requires only a little time and effort with the aid of a personal computer. The program I used took five minutes to write (I used MATLAB, but the code is so simple it is just as easy to do in any other language), ten minutes to type, and just one second to execute (on an 800-MHz machine). The result, after a total of $88{,}486$ floating-point arithmetic operations, is figure 3.4, which shows that θ does, indeed, have a rather broad maximum around $\theta = \theta_{\max} = 18.44°$, which occurs at $\alpha = 33.5°$.

3.3 The Envelope-Folding Problem

For our next problem, on how a computer can play a highly useful role in minimization analyses, consider the following problem that

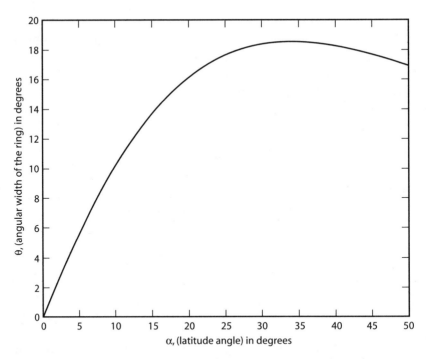

FIGURE 3.4. Observed angular width of Saturn's ring versus latitude.

appears to be deceptively simple. We are given a right triangle, with perpendicular sides of lengths a and b meeting at the corner O, as shown in figure 3.5. Suppose we fold the right angle over to place O at some point P on the hypotenuse. This can be done in an infinity of ways (with the folded triangle's sides OX and OY having lengths x and y, respectively, and such that $0 \leq x \leq a, 0 \leq y \leq b$). Each such way results in the folded triangle OYX having some area; our question is: what is the minimum possible area of OYX? This sounds like a simple question, but I don't think it is. If you don't agree, then shut the book right now and try your hand at it *before* you read what follows.

To start, let me make some elementary but crucial geometric observations. When we fold O onto P we create an *image* triangle (YPX, in dashed lines) that is a copy of the actual, folded triangle. For example, the angle YPX is a right angle because angle YOX is a right angle. Similarly, angle OYX equals angle PYX (called θ), and angle

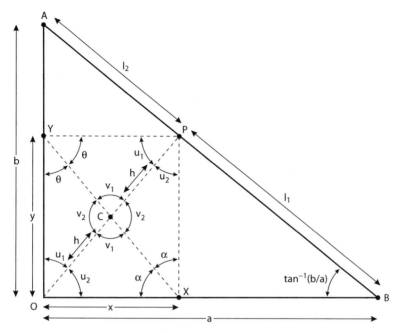

FIGURE 3.5. Geometry of the envelope-folding problem.

OXY equals angle *PXY* (called α). The other angles, called u_1, u_2, v_1, and v_2, are as shown in the figure. (Angle *YOP* $= u_1 =$ angle *YPO* because, by construction, triangle *YOP* is isosceles.) Finally, the dashed line segment *OP* (with length $2h$) is, of course, bisected by *YX* because the triangles *YPX* and *YOX* are identical. Most importantly, $v_1 = v_2 = 90°$, i.e., the line segment *OP* is perpendicular to the line segment *YX*. This last claim may or may not be obvious (try folding some paper triangles—that's what I did!), but it is easy to formally establish. That is,

$$\theta + u_1 + v_2 = 180° \text{ (triangle } OYC)$$

and

$$\theta + u_1 + v_1 = 180° \text{ (triangle } PYC),$$

and so $v_1 = v_2$. But, since $v_1 + v_2 = 180°$, we have $v_1 = v_2 = 90°$, as claimed. Also,

$$2\theta + 2u_1 = 180° \text{ (triangle } OYP)$$

and so $\theta + u_1 = 90°$, or $u_1 = 90° - \theta$. But as $u_1 + u_2 = 90°$, then $u_2 = 90° - u_1$, and so $u_2 = \theta$. All of this is straightforward, almost so simple you might wonder why I've bothered to spell it out. Here's why.

We have the area of the folded triangle as

$$A = \frac{1}{2} xy,$$

where

$$\frac{h}{x} = \cos(\theta) \quad \text{and} \quad \frac{h}{y} = \sin(\theta).$$

Thus,

$$A = \frac{h^2}{2\cos(\theta)\sin(\theta)}.$$

Next, using the law of cosines repeatedly on various triangles in figure 3.5, we can find h (and thus A) as a function of just θ.

In the notation of figure 3.5, we have

$$\ell_1^2 = (2h)^2 + a^2 - 2(2h)a\cos(\theta),$$

or

$$\boxed{\ell_1^2 = 4h^2 + a^2 - 4ha\cos(\theta).} \tag{1}$$

Also,

$$\ell_2^2 = (2h)^2 + b^2 - 2(2h)b\cos(90° - \theta),$$

or

$$\boxed{\ell_2^2 = 4h^2 + b^2 - 4hb\sin(\theta).} \tag{2}$$

And,

$$(2h)^2 = \ell_2^2 + b^2 - 2\ell_2 b \cos\left\{\tan^{-1}\left(\frac{a}{b}\right)\right\},$$

or, as

$$\cos\left\{\tan^{-1}\left(\frac{a}{b}\right)\right\} = \frac{b}{\sqrt{a^2 + b^2}},$$

we have

$$4h^2 = \ell_2^2 + b^2 - \frac{2\ell_2 b^2}{\sqrt{a^2 + b^2}}. \tag{3}$$

With one more application of the law of cosines, we have

$$(2h)^2 = \ell_1^2 + a^2 - 2\ell_1 a \cos\left\{\tan^{-1}\left(\frac{b}{a}\right)\right\},$$

or

$$4h^2 = \ell_1^2 + a^2 - \frac{2\ell_1 a^2}{\sqrt{a^2 + b^2}}. \tag{4}$$

Finally, we get our last equation from the Pythagorean theorem:

$$(\ell_1 + \ell_2)^2 = a^2 + b^2,$$

or

$$\ell_1 + \ell_2 = \sqrt{a^2 + b^2}. \tag{5}$$

Substituting (1) into (4) gives

$$\ell_1 = \frac{\sqrt{a^2 + b^2}}{2a^2}\left[2a^2 - 4ha\cos(\theta)\right],$$

while substituting (2) into (3) gives

$$\ell_2 = \frac{\sqrt{a^2 + b^2}}{2b^2} \left[2b^2 - 4hb \sin(\theta)\right].$$

Substituting these two results for ℓ_1 and ℓ_2 into (5) then gives us h as a function of θ:

$$h = \frac{1}{2\left[\dfrac{\cos(\theta)}{a} + \dfrac{\sin(\theta)}{b}\right]},$$

and so, at last, we have the area of the folded triangle as

$$A = \frac{(ab)^2}{8\cos(\theta)\sin(\theta)[a\sin(\theta) + b\cos(\theta)]^2}.$$

Setting $dA/d\theta = 0$ to find the minimum of A is a nasty business (try it!), and so I won't do that. Instead, let's use a computer to study the behavior of A directly.

We know, physically, that the extrema (minimum) of A occurs somewhere in the interval $0° \leq \theta \leq 90°$. Not all values of θ in this interval are possible, however, because we must satisfy the constraints of $0 \leq x \leq a$ and $0 \leq y \leq b$. The *maximum* value of θ occurs when we fold the entire length a up onto the hypotenuse (again, fold some actual paper triangles if this isn't clear). Thus,

$$2\theta_{max} + \tan^{-1}\left(\frac{b}{a}\right) = 180° \text{ (triangle } OPB\text{)},$$

or

$$\theta_{max} = 90° - \frac{1}{2}\tan^{-1}\left(\frac{b}{a}\right).$$

The *minimum* value of θ occurs when we fold the entire length b up onto the hypotenuse. Thus,

$$2(90° - \theta_{min}) + \tan^{-1}\left(\frac{a}{b}\right) = 180° \text{ (triangle } OPA\text{)},$$

or

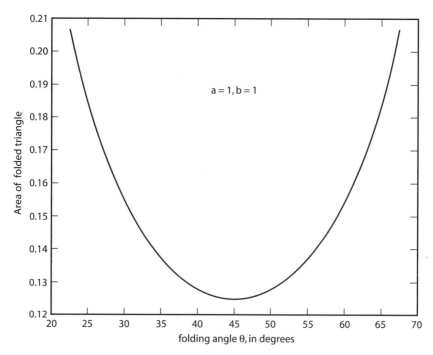

FIGURE 3.6. Area of the folded triangle, versus the folding angle $a = 1, b = 1$.

$$\theta_{\min} = \frac{1}{2} \, \tan^{-1}\left(\frac{a}{b}\right).$$

Figures 3.6 and 3.7 show $A(\theta)$ plotted over the interval $\theta_{\min} \leq \theta \leq \theta_{\max}$ for the cases of $a = b = 1$ and $a = 2, b = 1$, respectively. For the first case we get the obvious (by symmetry) result that $A_{\min} = 0.125$, at $\theta = 45°$, and in the second (not so obvious case) the answer is $A_{\min} = 0.2144$.

3.4 The Pipe-and-Corner Problem

Suppose we want to transport a long, cylindrical pipe through one underground tunnel (of width a) into another tunnel at a right angle to the first tunnel, all the while keeping the pipe horizontal. We imagine that during the move the pipe pivots on the tunnel corner

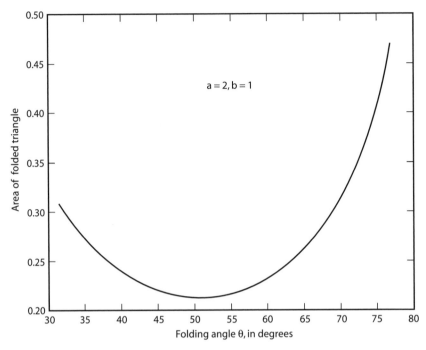

FIGURE 3.7. Area of the folded triangle, versus the folding angle $a = 2, b = 1$.

(point A) of figure 3.8, and also that it always slides along the left wall of the first tunnel (moving point B). Our question is: how wide must the second tunnel be to allow the pipe to be so moved? This is a popular textbook problem in introductory calculus courses, where it is invariably simplified by reducing the pipe's diameter to zero, i.e., by imagining in figure 3.8 that the pipe is a *line segment* of length ℓ and outside diameter $w = 0$. That makes it easy to find the maximum value of y, i.e., the maximum extension of the pipe across the second tunnel, which, of course, thus determines the required minimum width of the second tunnel for the move to be physically possible.

Far more realistic, however, is to allow the pipe to be able to have something inside of it, i.e., to have a nonzero diameter! From the geometry shown in figure 3.8, we have (in the notation of that figure)

$$d_1 + d_2 = \ell,$$

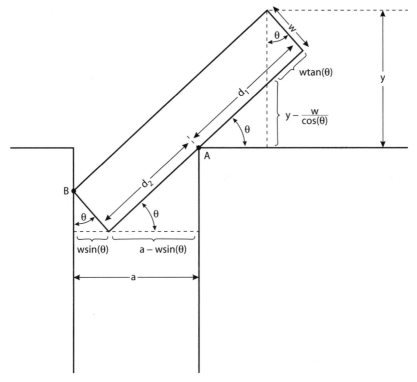

FIGURE 3.8. Geometry of the pipe-and-corner problem.

as well as

$$\frac{y - \dfrac{w}{\cos(\theta)}}{d_1 - w\tan(\theta)} = \sin(\theta)$$

and

$$\frac{a - w\sin(\theta)}{d_2} = \cos(\theta).$$

Solving these last two expressions for d_1 and d_2 and then substituting into the first expression, we can solve for y as a function of θ (the pivot angle):

$$y = \ell\sin(\theta) - a\tan(\theta) + \frac{w}{\cos(\theta)}.$$

It is physically obvious that there will be some unique angle $\theta = \hat{\theta}$, between 0° and 90°, at which y will attain its maximum value. To determine $\hat{\theta}$ analytically we could, as taught in freshman calculus (and as discussed in the next chapter), set the derivative of y with respect to θ equal to zero and solve for $\hat{\theta}$. If you try this, however, you'll get

$$\ell \cos^3(\hat{\theta}) - a + w \sin(\hat{\theta}) = 0,$$

which is *not* easily solved analytically for $\hat{\theta}$. We could, of course, just plot the left-hand side of this expression and observe where the curve crosses the θ-axis, and then plug that value for $\hat{\theta}$ back into the y-equation, but why bother? If we are going to use a computer to plot a curve then why not just use it to plot the y-equation itself and directly observe the maximum of y? And that's just what I'll do.

Since a is a "natural" dimension of the problem, let's actually study the so-called *normalized* equation

$$\frac{y}{a} = \left(\frac{\ell}{a}\right) \sin(\theta) - \tan(\theta) + \frac{\left(\frac{w}{a}\right)}{\cos(\theta)}.$$

That is, we will find the maximum of y *in units of a*, given both the pipe length and the pipe's outside diameter also in units of a. For example, if the pipe is 100 feet long, with an outside diameter of one foot, and if the first tunnel has a width of $a = 10$ feet, then our normalized equation becomes

$$\frac{y}{a} = 10 \sin(\theta) - \tan(\theta) + \frac{1}{10 \cos(\theta)}.$$

This equation is plotted in figure 3.9, which shows that y/a has a maximum value of 7.168. Thus, the second tunnel must be at least 71.68 feet wide.

If we had used the simple textbook model with $w = 0$, however, we would have calculated

$$\hat{\theta} = \cos^{-1}\left\{\sqrt[3]{\frac{a}{\ell}}\right\} = \cos^{-1}\left\{(0.1)^{1/3}\right\} = 1.0881 \text{ radians,}$$

which, when substituted back into the y-equation, gives the smaller result

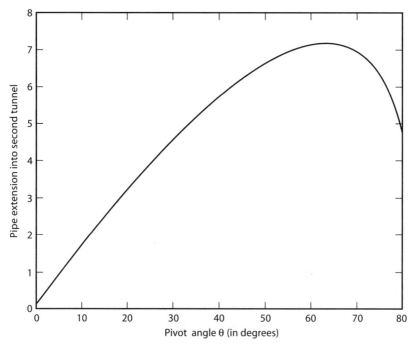

FIGURE 3.9. Turning a nonzero-diameter pipe around.

$$y_{max} = 100 \sin(\hat{\theta}) - \tan(\hat{\theta}) = 69.49 \text{ ft.}$$

Is this two-foot difference significant? Ask yourself that question the next time you try to move a (nonzero width) couch around a hallway corner from one room to another—half-an-inch (much less *two feet*) too little in the hallway width will ruin your day (I speak from experience!).

3.5 Regiomontanus Redux

Purists may not like the use of a computer to solve extremal problems, preferring pure mathematical demonstrations. They claim that while the sheer brute power of a modern computer may be sufficient to show some premise is either true or not true, such "calculate to exhaustion" demonstrations don't show *why* the conjecture is true or false. For example, the use of Eratosthenes' sieve to find the primes is perfect for use on a computer and yet nobody would claim to say it tells us, at some deep level, *why* there is an *infinity* of primes (and the sieve certainly doesn't tell us *anything* about the still open question of the infinity—or not—of the twin primes).

I expect that new generations of mathematicians will be able to expand their list of acceptable tools (which once included just the straight edge and the compass) to routinely include computers. Indeed, two famous extremal problems of mathematics have already yielded to computer analysis in the last quarter of the twentieth century. Thomas Hales (of honeycomb conjecture fame, mentioned in chapter 2) showed (in 1998, with the aid of enormous computer support) that the Kepler Sphere Packing Conjecture (dating from 1611) is true; face-centered cubic packing of identical spheres (the way oranges are displayed in pyramids in grocery stores) gives the maximum packing density. And Wolfgang Haken (1928–) and Kenneth Appel (born 1932 and now my colleague at the University of New Hampshire) at the University of Illinois showed (in 1976 and with the help of a huge computer program) that the Four-Color Conjecture (dating from 1852) is true: to color any planar map so that countries sharing a border have different colors requires, at most, four colors.

While the Regiomontanus problem, and its Saturn variant, are both clever and distinct in nature from the isoperimetric problems of the ancients, they too were initially treated with geometrical thinking. That was because the development of calculus, the next great step forward in the methods of extremal analysis, still had a century to wait. Many students today associate only the name of Newton with that development (or perhaps that of Leibniz as well, if they've heard a bit of history in their math or physics classes). In fact, it was the French lawyer and amateur mathematician Pierre de Fermat who took the first step toward introducing analytical techniques to extremal problems, where once only geometry was the means of attack. In the next chapter, then, Fermat and his work will take center stage. But, before Fermat, let's take one last look at the Regiomontanus problem and the use of a computer. The approach I'll use here is based on ideas presented in a letter by A. Tan and O. Castillo in the October 1983 issue of *Mathematics Teacher* ("Maximizing Paintings," p. 472).

The analysis of section 3.1 was literally one-dimensional, with the "painting" reduced to merely having a vertical dimension. A real painting, of course, also has a width, as shown in figure 3.10. Notice carefully that in that figure I have changed the symbols to be in agreement with Tan and Castillo. The painting's dimensions are now a and b, and the bottom edge of the painting is distance c above the eyes of the viewer. The viewer is imagined to be standing directly in front of the center of the painting, at a distance x from the vertical wall on which the painting is hanging.

As Tan and Castillo point out, what we *really* want to do is maximize the *solid* angle subtended at the viewer's eyes by the painting.

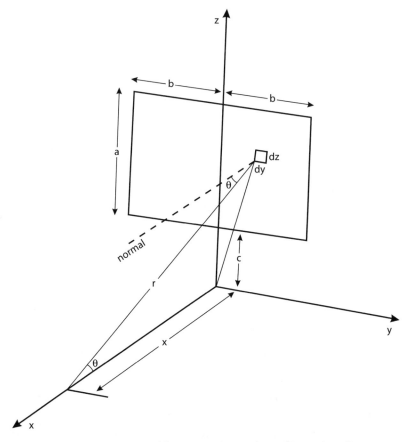

FIGURE 3.10. Regiomontanus' hanging picture (two-dimensional).

This will require us to evaluate a *double integral*, a big step beyond anything done so far in this book; you should definitely consider what follows, then, as optional reading (at least for now). If we locate an arbitrary patch of differential area of the painting, located at co-ordinates (y, z) as shown in figure 3.10, then that differential area is $dA = (dy)(dz)$ and the distance of dA from the viewer's eyes is, by a double application of the Pythagorean theorem, $r = (x^2+y^2+z^2)^{1/2}$. Now, if θ is the angle made by the viewer's line of sight to dA, then

$$\cos(\theta) = \frac{x}{r},$$

and the (differential) solid angle subtended at the viewer's eyes by dA is

$$d\Omega = \frac{dA\cos(\theta)}{r^2} = \frac{x\,dA}{r^3} = \frac{x\,(dy)(dz)}{\left(x^2 + y^2 + z^2\right)^{3/2}}.$$

We get the total solid angle subtended by the entire painting by integrating over all y and z that define the painting's extent, and so

$$\Omega = \iint_{\substack{\text{entire} \\ \text{painting}}} d\Omega = \int_{z=c}^{a+c} \int_{y=-b}^{b} \frac{x(dy)(dz)}{\left(x^2 + y^2 + z^2\right)^{3/2}}.$$

The actual details of doing the integrations are routine but lengthy and a bit messy (a good table of integrals is the "method" I used!), and so I'll simply quote the result:

$$\Omega = \sin^{-1}\left\{ \frac{(b^2 - x^2)\left[x^2 + (a+c)^2\right] - 2x^2b^2}{(b^2 + x^2)\left[x^2 + (a+c)^2\right]} \right\}$$

$$- \sin^{-1}\left\{ \frac{(b^2 - x^2)(x^2 + c^2) - 2x^2b^2}{(b^2 + x^2)(x^2 + c^2)} \right\}.$$

We could play with this, algebraically, to get a somewhat simpler appearing expression (indeed, Tan and Castillo's formula for Ω is a bit less intimidating), but I'm not going to bother. After all, what we would do next, analytically, would be to set the derivative of Ω with respect to x equal to zero and solve for x. Even with Tan and Castillo's expression (and certainly with mine) that proves to be an astonishingly ugly business! Even with their slightly less awful formula, Tan and Castillo were still forced to conclude "The value of x at which Ω maximizes can [read that as *must*!] be found numerically."

In figure 3.11, I have plotted Ω versus x for a particular Regiomontanus problem considered in section 3.1: a bug on the floor viewing a painting with its lower edge 8 feet above the floor and its upper edge 20 feet above the floor. (I first compared the results my expression gives for the x that maximizes Ω with the numerical results given by Tan and Castillo's expression for the examples treated in their analysis, and they agree exactly.) In the notation of figure 3.10, then, we have $a = 12$ feet and $c = 8$ feet. Assuming a *square* painting ($b = 6$ feet), the plot shows Ω is maximized when the bug is $x = 8.58$ feet from the wall, considerably closer than the value of $x = 12.65$ feet found when the painting was modeled as having only

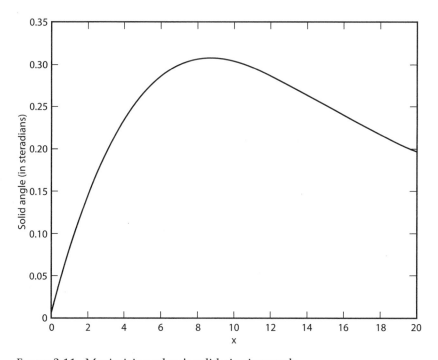

FIGURE 3.11. Maximizing a bug's solid viewing angle.

a vertical dimension. It is only for very wide paintings ($b \to \infty$) that the solid-angle solution approaches the solution for the case of the one-dimensional painting.

3.6   The Muddy Wheel Problem

For the final problem of this chapter, consider the geometry of figure 3.12, showing an event that a multitude of medieval mathematicians must have observed countless times: a wagon wheel rolling through a muddy street. The wheel, with radius R, is thickly coated with mud, and the rim is continually throwing off mud from every point. Our question here is: what is the maximum height above the ground reached by the ejected mud? The answer would be of considerable interest to those sitting in the wagon! To solve this problem, I'll use mathematical methods and physical arguments unknown to any medieval mathematician. How those methods came to be developed will be the central concern of the next two chapters; seeing

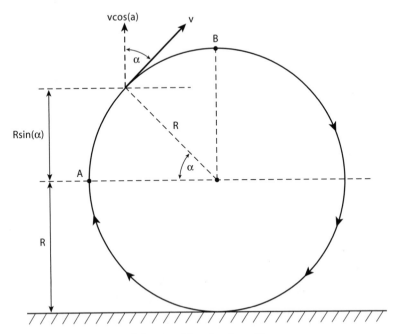

FIGURE 3.12. Geometry of the muddy wheel problem.

how neatly and elegantly these methods make short work of the muddy wheel problem will graphically illustrate how very different are the worlds of medieval and modern mathematics.

The maximum height of the tossed mud obviously depends on the speed at which the wheel rotates, and so let's say that all points on the rim are moving a steady speed of v. Thus, when mud comes off the rim, it is moving at speed v tangent to the rim, *at the instant it leaves the rim*. I think it is clear that, for the mud that reaches the maximum elevation above the road, it must come off the rim somewhere between the points marked A and B in the figure. So suppose, as shown, that the radius from the center of the wheel to some point in that quarter-circle makes angle α with the horizontal. The vertical component of the ejected mud's speed, at the instant of ejection, is thus $v \cos(\alpha)$; it is of course the *vertical* component *only* of the mud's speed that is responsible for the height reached by the mud. At the instant of ejection (let's call that instant time $t = 0$), the mud is therefore already at a height of $R + R \sin(\alpha)$ above the road.

Now, as the ejected mud rises, its vertical speed is continually reduced by gravity, which is given by $v \cos(\alpha) - gt$, where g is the acceleration of gravity. The mud's height *above the ejection point* is the integral of its speed, i.e., it is given by $v \cos(\alpha)t - \frac{1}{2}gt^2$. The mud reaches its maximum height, by definition, when its vertical speed has been reduced to zero. Thus, if $t = T$ is the time it takes to reach that maximum height, we have

$$v \cos(\alpha) - gT = 0,$$

or

$$T = \frac{v}{g} \cos(\alpha).$$

And so the maximum height *above the ground* reached by the mud, $h(\alpha)$, is

$$h(\alpha) = R + R \sin(\alpha) + v \cos(\alpha) \frac{v}{g} \cos(\alpha) - \frac{1}{2} g \frac{v^2}{g^2} \cos^2(\alpha),$$

or

$$h(\alpha) = R + R \sin(\alpha) + \frac{v^2}{2g} \cos^2(\alpha).$$

To find that α that maximizes $h(\alpha)$—let's call the maximum H—
we set $dh/d\alpha = 0$ and find that

$$R\cos(\alpha) - \frac{v^2}{g}\cos(\alpha)\sin(\alpha) = 0,$$

or

$$\sin(\alpha) = \frac{Rg}{v^2}.$$

This, of course, makes sense only if $Rg \leq v^2$ (because $|\sin(\alpha)| \leq 1$
for all real α) and so, for now, let's assume this condition is satisfied.
(I'll return to what $Rg > v^2$ means at the end of this section.) So,
since we also have

$$\cos^2(\alpha) = 1 - \sin^2(\alpha) = 1 - \frac{R^2 g^2}{v^4},$$

then

$$H = R + R\frac{Rg}{v^2} + \frac{v^2}{2g}\left(1 - \frac{R^2 g^2}{v^4}\right) = R + \frac{R^2 g}{v^2} + \frac{v^2}{2g} - \frac{R^2 g}{2v^2},$$

or

$$H = R + \frac{v^2}{2g} + \frac{R^2 g}{2v^2}.$$

Since we are assuming that $v^2 \geq Rg$, we see that our result says

$$H \geq R + \frac{Rg}{2g} + \frac{Rv^2}{2v^2} = 2R.$$

That is, as long as $v^2 \geq Rg$, then mud will always rise to at *least*
a height even with the top of the wheel (and, of course, the more
v^2 exceeds Rg, the more *above* the wheel will mud be flung, perhaps
onto the clothing of those riding too near to the sides of the wagon).
But what if $v^2 < Rg$? Then our condition of $\sin(\alpha) = Rg/v^2$ is
impossible to satisfy; so, let's return to the expression for $h(\alpha)$, to
just *before* we differentiated it. Then,

$$h(\alpha) = R\left[1 + \sin(\alpha) + \frac{v^2}{2Rg}\cos^2(\alpha)\right],$$

and so

$$h(\alpha) < R\left[1 + \sin(\alpha) + \frac{1}{2}\,\cos^2(\alpha)\right].$$

Defining $f(\alpha) = 1 + \sin(\alpha) + \frac{1}{2}\cos^2(\alpha)$, we have

$$h(\alpha) < Rf(\alpha).$$

Since

$$\frac{df}{d\alpha} = \cos(\alpha) - \cos(\alpha)\sin(\alpha) = \cos(\alpha)[1 - \sin(\alpha)],$$

and since $\sin(\alpha) \leq 1$ for all α, and since $\cos(\alpha) \geq 0$ for $0° \leq \alpha < 90°$ (α never *equals* 90°, as we have $v^2 < Rg$, not $v^2 \leq Rg$), then

$$\frac{df}{d\alpha} > 0, \qquad 0° \leq \alpha < 90°.$$

Since the derivative of $f(\alpha)$ is the slope of the tangent line to the $f(\alpha)$ versus α curve, then this result says $f(\alpha)$, over the semiclosed interval $0° \leq \alpha < 90°$, *approaches* a maximum as α *approaches* 90°. That is, over that interval $f(\alpha)$ *approaches* a maximum value of 2 as α approaches 90°. Thus, when $v^2 < Rg$, the value of H is strictly *less* than $2R$ (the top of the wheel). Mud coming off the wheel right at the top of the wheel (with no vertical speed) automatically achieves the height of $2R$, but certainly no mud is ever flung *above* the wheel if $v^2 < Rg$.

As late as the start of the seventeenth century, there were no mathematicians on earth who could have done this analysis. At the end of the seventeenth century, there were many. What happened during that century—that advanced the mathematics of extrema in such a revolutionary way—is the central topic we take up next.

Solution to the Problem in Section 3.1

The answer to the trousers/mirror version of the Regiomontanus problem is that the man should stand at a distance of $\frac{1}{2}\sqrt{h(h-\ell)}$ from the mirror. (The key observation is that the

(continued)

man's trousers appear as far *behind* the mirror as he stands in front of it). This *is* the correct mathematical result, but does it really "make sense"? For example, I wear 31-inch trousers and my eyes are 70 inches above the floor. But, when I try trousers on at my local men's clothing store, I stand significantly farther away from the dressing room mirror than $\frac{1}{2}\sqrt{70(70-31)}$ = 26.1 inches. Apparently simply maximizing the viewing angle does not really capture what is meant by "best view."

4.

The Forgotten War of
Descartes and Fermat

4.1 Two Very Different Men

Modern students, when first introduced to the differential calculus, learn that it was the simultaneous and independent creation of the Englishman Isaac Newton (1642–1727) and the German Gottfried Leibniz (1646–1716). Perhaps they are told that Newton and Leibniz (and their respective followers) engaged in a lengthy and acrimonious debate over intellectual priority, and that Newton continued the battle even after Leibniz's death, right up to the day of his own death. Almost certainly, however, they are not told anything about an equally nasty war of words between two French mathematicians a half-century earlier, on some of the same issues that later engaged Newton and Leibniz.

Pierre de Fermat (1601–65) and René Descartes (1596–1650) were very different men. Fermat was a family man, trained as a lawyer who loved mathematics as a pastime, and who so valued his privacy that he published very little (and even then, only anonymously). He was a classical scholar as well, fluent in Italian, Spanish, Latin, and Greek, and an omnivorous student of the writings of the ancient mathematicians. Descartes, also trained as a lawyer, was a man who soon came to embrace public acclaim, who published widely, and who devoted his whole life to the single-minded pursuit of abstract knowledge. A family would have been a distraction, and Descartes never married (although in 1635 he did have a daughter by one of his servants).

Unlike Fermat, Descartes gave the impression that he was often uninformed of what others had done before him; at least he only rarely mentioned the work of anybody else in his writings. And when he did, it was often in the most unpleasant manner one could imagine: at various times in his life he called his critics "two or three flies," "less than a rational animal," "a little dog," and "extremely contemptible." The actual works of others were often rejected in incredibly offensive language, e.g., as being fit only for use as "toilet paper" or, in the case of Fermat, as being "shit."

We remember both men for very different reasons than what they fought over: Descartes for his philosophical writings and the joining of algebra with geometry into analytic geometry, and Fermat for his work in probability and number theory, particularly the famous and only recently resolved "Fermat's Last Theorem." What these two brilliant intellects battled over, however, was none of this, but rather first a problem in physics, and then the beginnings of how to answer extremal questions through analysis rather than the classical tool of geometry.

The origins of the conflict between the two men can be found in Descartes' essay on optics, *La Dioptrique*, one of the appendices in his 1637 book *Discourse on Method*. There he treated the phenomenon of the refraction of light, which is the next natural question to pursue after noticing the details of the reflection of light. (Descartes' interest in the law of refraction was also motivated by his research into the nature of the rainbow, which he—and others before him—correctly believed to be due to the scattering of sunlight by water droplets in the air. Descartes needed the law of refraction to mathematically describe that scattering, and I'll return to the rainbow problem in the next chapter.) Both phenomena, reflection and refraction, involve extremal arguments of a quite different nature (and the solution to one of the first problems in the calculus of variations—discussed in chapter 6—used the refraction law), so let me make a brief digression to describe them.

"And God said, 'Let there be light'; and there was light. . . . But we can imagine the angelic architect asking for more details: 'What path shall light follow in going from P to Q?' And the answer might have been, 'Don't bother me with such details. See that it makes the trip in minimum time.' From this *minimal* principle one finds that *for reflection* the angle of incidence should equal the angle of reflection, while *for refraction* at an interface the ratio of the sine of the angle of incidence to the sine of the angle of refraction must equal the ratio of speeds in the two media."

"And God saw that the light was good."
　　　　　—Arthur Bernhart, *Scripta Mathematica* 1959, p. 206.

Professor Bernhart might have mentioned, however, that there are *two* ways to form the ratio of the speeds; Descartes got it wrong, but Fermat got it right as you'll see in what follows.

4.2　Snell's Law

It was Euclid who first made note (three centuries before Christ) of the now familiar reflection law of light: if a beam of light is sent toward a mirror, then the angle of incidence equals the angle of reflection ($\theta_i = \theta_r$ in figure 4.1), not only for a flat mirror but for

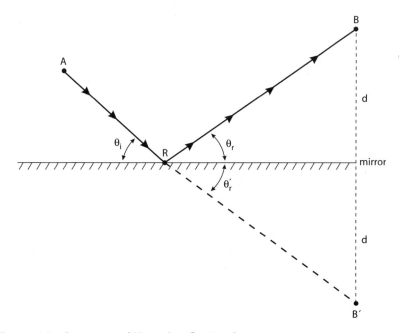

FIGURE 4.1. Geometry of Heron's reflection law.

a curved mirror as well (for curved mirrors we measure the two angles with respect to the tangent line at the point of reflection, R). It was Heron of Alexandria, however, who first observed (in the first century A.D., in his book on mirrors, *Catoptrica*) that the reflection law is the immediate consequence of assuming that the beam path ARB is the *minimum length path*. That is, if the point R on the mirror were such that $\theta_i \neq \theta_r$, then the resulting total path length would be longer. (The implicit assumption is, of course, that the beam of light *does* reflect off of the mirror—the absolute shortest path from A to B is simply the direct, straight line segment joining the two points. Indeed, if a light bulb is at A, broadcasting light in all directions, then B receives light along the *two* paths ARB and AB.) Heron's observation is the first occurrence of a *minimum principle* in mathematical physics; such principles play central roles in modern theoretical physics. It is impressive and instructive to examine how Heron derived the reflection law from this particular minimum principle.

If the destination point B is distance d above the mirror, then B's reflected point (B') is distance d "below" the mirror. RB and RB' are, therefore, the equal-length hypotenuses of two congruent right triangles, which means $\theta_r' = \theta_r$ (referring again to figure 4.1). Now, the total light path length is $AR + RB = AR + RB'$, and this last sum is the path length from A to B'. The shortest path from A to B' (and so the shortest length for the reflected path, as well) is along a straight line, and so $\theta_r' = \theta_i$, which immediately gives $\theta_i = \theta_r$, i.e., the reflection law.

With the reflection law thus established, attention turned next to *refraction*, the phenomenon of the change in direction experienced by a beam of light when it crosses the interface between one transparent medium into another (from air into water or into glass, for example), as shown in figure 4.2. Attempts to formulate a mathematical description of refraction can be traced as far back as Ptolemy; the first preliminary mathematical results appeared in the German astronomer Johannes Kepler's 1611 *Dioptrice*, but it wasn't until the experimental work of the Dutch physicist Willebrord Snel (1580–1626) that the precise form of the refraction law was discovered. If we measure the angles θ_i and θ_r with respect to the interface normal (the dashed line in figure 4.2), then Snel discovered around

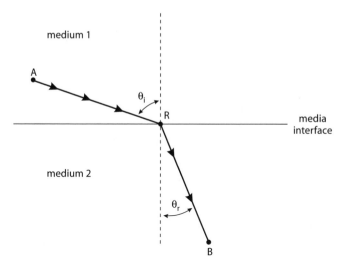

FIGURE 4.2. Geometry of Snell's refraction law.

1621 (but didn't publish) what is now called Snell's law (the double-l spelling is from the Latinized form of his name, Snellius):

$$\frac{\sin(\theta_i)}{\sin(\theta_r)} = \text{"constant"},$$

where the "constant" is a function of the nature of the two media. Snel observed that if medium 2 is denser than medium 1 (as with a light beam traveling from air into water) then the "constant" is greater than one. That is, $\sin(\theta_i) > \sin(\theta_r)$ or, equivalently, $\theta_i > \theta_r$; i.e., upon entering the water the light beam bends *toward* the normal.

Snel's notes on his experiments were lost some time after 1662, and what we know of them is only through the writings of those who saw them. Somewhat less well known, unpublished experimental research leading to Snell's law, years before Snel, is also attributed to the English mathematical physicist Thomas Hariot (1560–1621). Hariot died from a cancer of the nose—due to a youthful intoxication with tobacco while serving as the science officer on a colonizing expedition to Virginia in 1585(?)—and his scientific research had ceased by 1618, three years before Snel's research. History records, however, that it was Descartes who first published, in *La Dioptrique*, a theoretical derivation of the law of refraction, as well as offering an explanation of the "constant." Many historians have long believed that Descartes learned of Snel's experimental work and then used that knowledge to guide the often strained physical assumptions of the nature of light that appear in his analysis. Other historians disagree but, for a while at the end of the 1600's, there were nasty rumblings in the scientific community about plagiarism on Descartes' part. Descartes' former admirer, Christiaan Huygens, who as a young boy met Descartes often when the Frenchman visited Huygens' father, was among those who suspected the worse. The matter is still not fully resolved.

It is Descartes' "derivation" of Snell's law that Fermat read in 1637 and found lacking in merit, and he said as much in a letter to a correspondent who also had contact with Descartes. Fermat's skeptical reaction soon got back to Descartes, and the war was on. So, how *did* Descartes derive Snell's law, and why did Fermat think that derivation wrong, a view shared by all modern physicists since the middle of the nineteenth century?

Adopting a particle view of light, Descartes began his analysis by making an analogy with a tennis ball hitting a cloth barrier at incident angle θ_i. He argued that the ball would lose some of its speed in the vertical direction *only*, because the cloth would offer no resistance to the ball in the horizontal direction. The horizontal speed component, therefore, would be unchanged. To express this claim mathematically, let the ball's speed before hitting the cloth barrier be v_1 (as shown in figure 4.3) and v_2 after penetrating the cloth. Then, the ball's horizontal component of speed above the cloth (in medium 1) is $v_1 \sin(\theta_i)$, which is, according to Descartes, also the horizontal component of speed below the cloth (in medium 2). Since that component is $v_2 \sin(\theta_r)$, then

$$v_1 \sin(\theta_i) = v_2 \sin(\theta_r),$$

or

$$\frac{\sin(\theta_i)}{\sin(\theta_r)} = \text{constant} = \frac{v_2}{v_1}.$$

The first equality is Snell's law (the ratio of sines is a constant), but Descartes' derivation has actually led him into serious difficulty. Snel

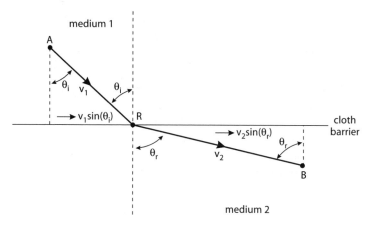

FIGURE 4.3. Geometry of Descartes' refraction law "derivation."

had not explained what the constant actually is, while Descartes had apparently shown it is the ratio of the ball's speeds in the two media. If we stick with the tennis ball analogy to light (i.e., a particle or so-called *corpuscular* view of light), with $v_2 < v_1$ (the ball loses speed as it penetrates the cloth barrier), then Descartes' version of Snell's law must be wrong since it says

$$\frac{\sin(\theta_i)}{\sin(\theta_r)} < 1, \qquad \text{i.e., } \theta_i < \theta_r.$$

That is, the ball (particle of light) would veer *away* from the normal as it enters a denser medium. As mentioned before, however, light is observed to do precisely the opposite. To bring his result into agreement with experiment, then, Descartes had to drop the ball analogy in midstream and conclude that $v_2 > v_1$, i.e., that light *speeds up* as it penetrates the barrier. Thus, Descartes was forced to conclude that the speed of light is greater in denser media. In Descartes' day there was no experimental measurement of the speed of light in *any* medium, and so no one could say whether he was right or not. And experiment is, of course, the only way to really settle such a question—unless you are a philosopher. To "explain" how light gains speed as it passes from air into water Descartes put forth arguments in the tradition of Aristotle ("physics the way we *think* it should be, rather than what experiment *shows* it to be"), arguments that today seem ludicrous.

Descartes' view of space was that there is no *empty* space (for him a vacuum was impossible), and that all apparently empty space between macroscopic bodies was actually filled with an invisible "something." This view was long in dying, and right up to the end of the nineteenth century all physicists—until Einstein—believed that light needed that "something" to move through. They called it the *ether* and, while not Descartes' term, it represented his view. Descartes also believed that light was a "pressure" that traveled infinitely fast (some historians claim he really only asserted it travels *very* fast, not infinitely fast) through the "something," a view that would be logically at odds with an assertion that light travels *faster* in water than in air. Both Fermat and Descartes were dead before the first experimental measurement of the *finite* speed of light was made (an astronomical experiment in 1675 by the Danish astronomer Olaus Roemer, based on the timing of Jupiter's eclipsing of its moons). And it wasn't until almost another two *centuries* later that the speed of light in water was measured to be less than that in air (a terrestrial experiment in 1850, by the French physicists Hippolyte Fizeau and Jean Foucault). If one assumes that light is the wave phenomenon pioneered by Huygens, rather than a particle one, then the slowing of light in water allows one to derive Snell's law of refraction (see any college freshman physics textbook); indeed, the Fizeau/Foucault result was interpreted as proof that beams of light are waves, not particles. Things are actually not quite that simple, but that is an issue for a book on quantum electrodynamics!

You can find more on Descartes and his flawed optical physics, at a fairly technical level, in the paper by W. B. Joyce and Alice Joyce, "Descartes, Newton, and Snell's Law" (*Journal of the Optical Society of America*, January 1976, pp. 1–8), and at the historian's level in the books by William R. Shea, *The Magic of Numbers and Motion: The Scientific Career of René Descartes* (Science History Publications 1991) and A. I. Sabra, *Theories of Light: From Descartes to Newton*, (Cambridge University Press 1981). The second book, in particular, details the bitter feelings Descartes had toward Fermat because of Fermat's rejection of Descartes' "derivation" of Snell's law. You can find more on Descartes' flawed physics, in general, in Herman Erlichson's "The Young Huygens Solves the Problem of Elastic Collisions" (*American Journal of Physics*, February 1997, pp. 149–54). And finally, you can find a very detailed description of Hariot's ingenious experiments on refraction in John W. Shirley, "An Early Experimental Determination of Snell's Law" (*American Journal of Physics*, December 1951, pp. 507–8).

When Fermat read *La Dioptrique* he was unimpressed and, as mentioned earlier, was blunt in his criticism. He wrote, in part, "of all the infinite ways [to analyze the motion of light] the author [Descartes] has taken only that one which serves him for his conclusion; he has thereby accommodated his means to his end, and we know as little about the subject as we did before." An uncharitable reading of this might be that Descartes knew what the answer must be—from his earlier knowledge of Snel's experimental work—and so he simply fiddled with his physical assumptions until he got what he knew experiment said he *had* to get. In other words, Descartes' so-called derivation of Snell's law of refraction was no more than a begging of the question. (When I was a college undergraduate, the writing of a made-up lab report for a missed chemistry experiment was called a "dry lab," as compared to actually *doing* the experiment and getting real data, which was, of course, a "wet lab." Fermat thought Descartes' "derivation" to be a dry lab!) Further, Fermat rejected as nonsense Descartes' assertion of the infinite speed of light and his subsequent illogical argument that light travels faster (than infinity?) in water than in air. Fermat's position was that light traveled at a (very fast) *finite* speed in air, and that it was slowed when traveling through a denser ("more resistive") medium such as water.

Fermat initially believed that, since Descartes' derivation was clearly (to Fermat) built on sand, then the "ratio of sines is a constant" result must be incorrect. Eventually Fermat learned that the formula was, in fact, generally accepted as true because it could be verified by direct experiment! This greatly puzzled Fermat; how had Descartes managed to derive the correct law of refraction from erroneous arguments? It became a quest for Fermat to find a physically correct derivation of the law of refraction; he believed that the law would be mathematically different from Descartes' ratio of sines result while also being able to give nearly the same *numerical* results, thus explaining the (coincidental) experimental agreement with Descartes' result. With Fermat's subsequent great discovery of the "principle of least time" (discussed later in this chapter) his quest ended in 1658 with success—but with a surprising twist that astonished Fermat.

To carry out the calculations involved in applying the principle of least time to the refraction of light, Fermat used new mathematical techniques of his own devising, including *almost* what is now called the *derivative* of a function. Stimulated by an observation due to Kepler (see the box at the end of this section)—at an extrema, either minimum or maximum, a function $f(x)$ is not changing as tiny changes are made in x—Fermat transformed Kepler's insight into mathematics. Completed by 1629, Fermat published his discoveries in 1637 as *Method for Determining Maxima and Minima and Tangents to Curved Lines*. The date is important, as Descartes saw it just after learning of Fermat's rejection of *La Dioptrique*, and so, Descartes being Descartes, replied in kind to Fermat's work. (Descartes also saw a challenge in Fermat's *Method* to yet another of Descartes' appendices to his *Discourse*; Descartes thought his *Geometry* did what Fermat claimed to do, only better, with his—Descartes'—mathematics.)

The irony in all of this is delicious. Descartes rejected Fermat's work on maxima and minima largely because Fermat had rejected Descartes' derivation of the law of refraction. Then Fermat used his maxima/minima technique to correctly derive the refraction law, as well as giving a proper explanation to the "constant" in Snell's law. We'll take up Fermat's mathematics for our next discussion.

When engineers and scientists think of Johannes Kepler (1571–1603) it is almost certainly in connection with his famous three laws of planetary motion. An often ignored aspect of his genius, however, is his contribution to the early development of the differential and integral calculus. What is particularly amusing about this is what motivated Kepler in those mathematical researches; shortly after his second marriage in 1613, while setting up a new household, he learned how wine merchants determined the "volume" of wine barrels. They simply stuck a rod in through a hole at the edge of the top lid and measured the length of the barrel diagonal from top to bottom, without regard to the actual shape of the barrel. This made no sense to a man with Kepler's mathematical ability, of course, and he began to think upon the question of just how one would compute the volumes of various barrel shapes.

Kepler published the results of his work in the 1615 book *Stereo-metria doliorum vinariorum (New Solid Geometry of Wine Barrels)*. One result is particularly interesting for us: of all cylinders with the same diagonal, the one with the maximum volume is the one in which the ratio of the diameter to the height is $\sqrt{2}$ (a rather squat barrel resembling, in fact, the storage tanks used to hold oil in petroleum refineries). This result is worked out in the next chapter as an example of the new calculus of Newton and Leibniz.

4.3 Fermat, Tangent Lines, and Extrema

Recall the problem we solved back in chapter 1, using the method of completing the square: how should a constant C be divided into two parts so their product is maximized? There we wrote the two parts as x and $C - x$, and their product as $M = x(C - x)$. Fermat solved this same problem with a new approach, as follows. Expanding, we have

$$x^2 - Cx + M = 0.$$

Solving for x gives

$$x = \frac{C \pm \sqrt{C^2 - 4M}}{2}.$$

He next argued that if M *is* the maximum product then there should be just one value of x that achieves that maximum. Thus, the quantity under the square root sign must be zero. So,

$$M = \frac{1}{4}C^2$$

and

$$x^2 - Cx + \frac{1}{4}C^2 = 0 = \left(x - \frac{1}{2}C\right)^2 = 0.$$

Thus, $x = \frac{1}{2}C$ gives the maximum product.

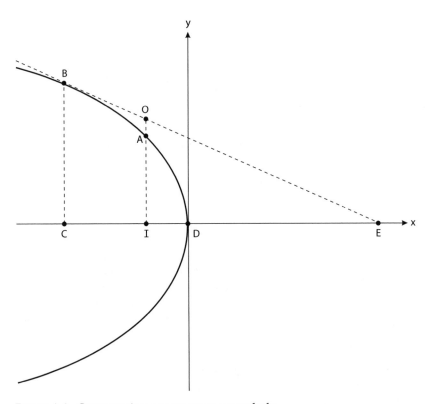

FIGURE 4.4. Constructing a tangent to a parabola.

Fermat also applied his "double-root" idea to the problem of drawing tangents to a given curve, at a given point. For example, consider the parabola $x = -y^2$ shown in figure 4.4. Suppose the given point is B, with coordinates $x = -s, y = \sqrt{s}, s \geq 0$. The generic equation of the tangent line is, of course, the well-known equation for a straight line

$$y = mx + b,$$

where m is the slope and b is a constant. Now, this straight line intersects the parabola at the solutions to

$$x = -(mx + b)^2,$$

which is easily solved to give

$$x = \frac{-(2mb + 1) \pm \sqrt{(2mb + 1)^2 - 4m^2b^2}}{2m^2}.$$

Fermat next invoked his central argument, that there is only *one* actual intersection of the *tangent* line with the curve, and so it must be true that

$$(2mb + 1)^2 - 4m^2b^2 = 0.$$

This is easily solved to give

$$b = -\frac{1}{4m}.$$

Therefore, at the given point B we have (since B is on the tangent line)

$$\sqrt{s} = -ms - \frac{1}{4m},$$

which is again easily solved to give

$$m = -\frac{1}{2\sqrt{s}} \quad \left(\text{and so } b = \frac{1}{2}\sqrt{s}\right).$$

Thus, the equation of the tangent line is

$$y = -\frac{1}{2\sqrt{s}} x + \frac{1}{2}\sqrt{s}, \qquad s \geq 0.$$

To actually draw the tangent line, it is sufficient to locate point E in figure 4.4, the intersection point of the tangent line with the x-axis. Setting $y = 0$ then, the x-coordinate of E is $x = s$. So, Fermat's procedure for drawing the tangent line at any given point B on the parabola $x = -y^2$ is the following four-step process:

1. drop the perpendicular from B to the x-axis, to the point C in figure 4.4.
2. measure the length $CD = s$, where D is the coordinate system origin.
3. $DE = s$, too, thus determining E.
4. connect B and E with a straight line.

Fermat later altered his "double-root" argument into what is nearly the modern approach for finding the extrema of a function by setting the first derivative to zero. That is, returning to the example that started this section, suppose $\hat{x}$ is the value of x that gives the maximum product, and that E is a "very small" quantity. Then using $\hat{x}$ or $\hat{x} + E$ should give nearly equal results, i.e.,

$$\hat{x}^2 - C\hat{x} + M \approx (\hat{x} + E)^2 - C(\hat{x} + E) + M,$$

and the near-equality will become a true equality as we let $E \to 0$. So, expanding and canceling equal terms on both sides, we arrive at

$$0 \approx 2\hat{x}E + E^2 - CE.$$

Since we haven't yet let E go all the way to zero, we can divide through by E to get

$$0 \approx 2\hat{x} + E - C.$$

This division by E is crucial, of course, because otherwise as we let $E \to 0$ we would get nothing but the undeniably true (but not very interesting and certainly not useful) tautology of $0 = 0$. But, if *after* the division we let $E \to 0$, we get the equality

$$0 = 2\hat{x} - C.$$

Thus, as before, $\hat{x} = \frac{1}{2}C$.

As another example of this technique, Fermat showed how to find the extreme value of the more complicated function $f(x) = ax^2 - x^3$, where a is a given constant. As before, let $\hat{x}$ be the value of x that gives the extreme value of f, and let E be "very small." Then,

$$f(\hat{x}) \approx f(\hat{x} + E),$$

or

$$f(\hat{x}) - f(\hat{x} + E) \approx 0,$$

with the near-equality becoming an equality as $E \to 0$. So,

$$\left[a\hat{x}^2 - \hat{x}^3\right] - \left[a(\hat{x} + E)^2 - (\hat{x} + E)^3\right] \approx 0,$$

or, after expanding and canceling, and dividing through by E,

$$-2a\hat{x} - aE^2 + 3\hat{x}^2 + 3\hat{x}E + E^2 \approx 0.$$

Then, letting $E \to 0$, the near-equality becomes an equality and

$$-2a\hat{x} + 3\hat{x}^2 = 0 = \hat{x}(-2a + 3\hat{x}).$$

There are *two* solutions: $\hat{x} = 0$ and $\hat{x} = 2a/3$. Notice that

$$f(\hat{x} = 0) = 0$$

and

$$f\left(\hat{x} = \frac{2a}{3}\right) = \frac{4a^3}{27}.$$

So, with reference to figure 4.5, what these results say is that $\hat{x} = 2a/3$ gives a *local maximum* if $a > 0$ (in the left plot, where $a = 3$ and $\hat{x} = 2$) and a *local minimum* if $a < 0$ (in the right plot, where $a = -3$ and $\hat{x} = -2$). The $\hat{x} = 0$ solution gives a *local* minimum if $a > 0$ and a *local* maximum if $a < 0$. For both $\hat{x}$ values, no matter what the sign of a, there is no absolute or *global* minimum *or* maximum since $f(x)$ becomes unbounded as $x \to \pm\infty$.

The reason for emphasizing that the extremas of $f(x)$ are *local* is that extrema are completely distinguished by the behavior of the function in the *neighborhood* of the extrema. It is entirely possible, for example, to have a two-extrema function with its local minimum *larger* than its local maximum. An example of this is shown in figure 4.6, for the function $f(x) = x + (1/x)$ (which is, of course, discontinuous at $x = 0$). The local minimum at $x = +1(f(+1) = 2)$ is larger than the local maximum at $x = -1(f(-1) = -2)$.

As a much more complicated example of his method, Fermat also treated a geometric problem from Pappus' *Mathematical Collection* (Proposition 61), one that leads (in modern algebraic notation) to calculating the minimum of a ratio of two polynomials:

$$\frac{(a - x)(b + x)}{x(c - x)},$$

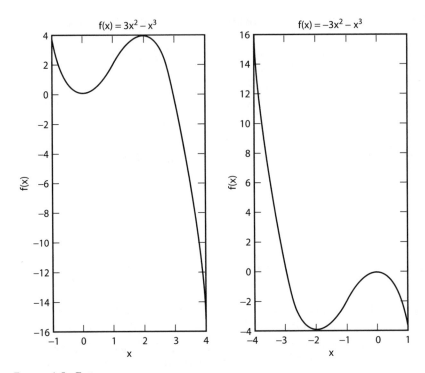

FIGURE 4.5. Extrema.

where a, b, and c are given constants. This is not an easy problem! You can find the original geometric statement in Alexander Jones' translation of Book 7 of the *Collection* (Springer-Verlag 1986, p. 186.)

4.4   The Birth of the Derivative

It is obvious, at this point, that in his examples Fermat was essentially calculating the *limit* that we call today the first derivative,

$$\lim_{E \to 0} \frac{f(x+E) - f(x)}{E} = \frac{df}{dx} = f'(x),$$

and then setting it equal to zero. (Today's textbooks commonly use ε, or Δx, rather than E, in this definition.) This definition was not formally introduced into mathematics until 1817, by the Czech

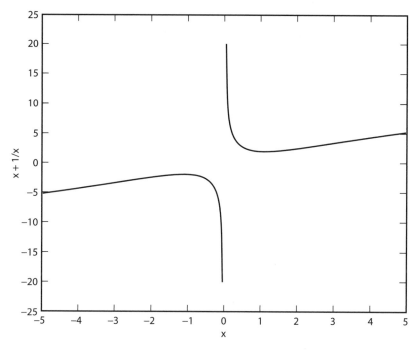

FIGURE 4.6. A function with relative minimum > relative maximum.

mathematician Bernard Bolzano (1781–1848), but the *idea* was in Fermat's work long before 1817. Indeed, it was in print in Fermat's 1637 *Method*, five years before Newton was born and nine years before Leibniz's birth (the two men normally credited with the invention of the differential calculus). This tells us, for example, that the derivative of any constant is zero, since $df = 0$ (*constants* don't change!) There were, of course, others who also contributed to the concept of the derivative, e.g., the Dutch mathematician Johann Hudde (1628–1704), who in 1659 showed how to differentiate a polynomial of any degree and so how to find its extrema. An outstanding historical exposition on the evolution of the derivative is the paper by Judith V. Grabiner, "The Changing Concept of Change: The Derivative from Fermat to Weierstrass" (*Mathematics Magazine*, September 1983, pp. 195–206). Grabiner starts off with the wonderful observation "The derivative was first *used*; it was then *discovered*; it was then *explored and developed*; and it was finally *defined*."

As a more sophisticated example of the use of the limit definition of the derivative, which will provide us with a result we'll use in the next section, suppose we know how to differentiate some simple function of x, called $g(x)$. For example, if $g(x) = x$ or x^2, then the derivative is 1 or $2x$, respectively (both results follow *easily* from the derivative definition—try it!). Then, our question is: what is the derivative of $h(x) = \sqrt{c + g(x)}$, where c is a constant?

Using the definition of the derivative, we have

$$\frac{dh}{dx} = \lim_{\varepsilon \to 0} \frac{h(x + \varepsilon) - h(x)}{\varepsilon} = \lim_{\varepsilon \to 0} \frac{\sqrt{c + g(x + \varepsilon)} - \sqrt{c + g(x)}}{\varepsilon}.$$

Now,

$$\frac{dg}{dx} = \lim_{\varepsilon \to 0} \frac{g(x + \varepsilon) - g(x)}{\varepsilon},$$

or, approximately, if ε is not equal to zero but "very close" to zero:

$$\varepsilon \frac{dg}{dx} + g(x) \approx g(x + \varepsilon).$$

Thus,

$$\frac{dh}{dx} = \lim_{\varepsilon \to 0} \frac{\sqrt{c + g(x) + \varepsilon \dfrac{dg}{dx}} - \sqrt{c + g(x)}}{\varepsilon}$$

$$= \lim_{\varepsilon \to 0} \frac{\sqrt{\{c + g(x)\} \left\{ 1 + \dfrac{\varepsilon}{c + g(x)} \dfrac{dg}{dx} \right\}} - \sqrt{c + g(x)}}{\varepsilon}$$

$$= \sqrt{c + g(x)} \lim_{\varepsilon \to 0} \frac{\sqrt{1 + \dfrac{\varepsilon}{c + g(x)} \dfrac{dg}{dx}} - 1}{\varepsilon}.$$

Next, and finally, using the approximation $\sqrt{1 + u} \approx 1 + \frac{1}{2}u$ for u "small," we have our result:

$$\frac{dh}{dx} = \sqrt{c + g(x)} \lim_{\varepsilon \to 0} \frac{1 + \dfrac{1}{2} \dfrac{\varepsilon}{c + g(x)} \dfrac{dg}{dx} - 1}{\varepsilon}$$

$$= \sqrt{c + g(x)}\, \frac{1}{2} \cdot \frac{1}{c + g(x)} \frac{dg}{dx} = \frac{1}{2} \cdot \frac{1}{\sqrt{c + g(x)}} \frac{dg}{dx}.$$

The modern notation for the derivative, e.g., dx/dt and d^2x/dt^2 for the first and second derivatives of $x(t)$ with respect to t (time), is due to Leibniz. In Newton's notation they would be written as $\dot{x}(t)$ and $\ddot{x}(t)$, respectively. Newton's dot notation is still used today, but is generally regarded as less useful. Leibniz's notation lends itself to the useful device of thinking of the differentials dx and dt as algebraic quantities, and to treating them as such. For example, a little later in the next section I'll formally derive what is called the *chain rule*, but in Leibniz's notation (and *not* in Newton's) it is trivially obvious: if $u(t)$ and $v(t)$ are two functions of the independent variable t, and if $f(t) = u\{v(t)\}$, then

$$\frac{df}{dt} = \frac{du}{dv} \cdot \frac{dv}{dt}$$

"because" we can cancel the two dv differentials on the right-hand side.

Even Newton's name for the new math has been discarded. Finding his original motivation in considering how quantities change with the "flux of time" (in his *Principia* he writes of time as *flowing*), Newton called $x(t)$ a flowing quantity, or *fluent*, and the rate at which $x(t)$ changes with time (that is, the derivative of $x(t)$) the *fluxion*. The use of the word *calculus*, rather than Newton's "method of fluxions," is again due to Leibniz from some time before 1680. Newton himself had adopted Leibniz's term by 1691.

Much of the failure by Fermat to receive credit for his wonderful discovery is almost certainly due to the criticisms of Descartes, who simply failed to appreciate what he read in *Method*. This isn't to say all mathematicians failed to appreciate Fermat's contributions to the invention of the differential calculus. The Italian-born French mathematician Joseph Lagrange (1736–1813), who developed the modern approach to the calculus of variations (see chapter 6), wrote "One may regard Fermat as the first inventor of the new calculus." And the French genius Pierre Simon de Laplace (1749–1827) declared "Fermat should be regarded, then, as the true discoverer of Differential Calculus." Modern historians disagree, however, arguing that Fermat's calculations are quite limited in scope, while Newton and Leibniz developed calculus in breadth; in particular, they discovered *general* formulas for the differentiation of complicated functions. Still, when Lewis Trenchard More published his 1934 biography *Isaac Newton* (Charles Scribner's Sons) he announced (p. 185) that he had discovered, in the major archival holdings of Newton's papers, a previously unknown draft of a letter in which Newton himself stated his debt to Fermat for the invention of the differential calculus: "I had the hint of this method from Fermat's way of drawing tangents and by applying it to abstract equations, directly and indirectly, I made it general."

   To see why Newton wrote those words, consider again the parabola $x = -y^2$ shown in figure 4.4. Recall that Fermat took the tangent at B as the dashed line through B that intersects the x-axis at E. Dropping the perpendicular from B to the x-axis (to C), then, reduces the problem of drawing the tangent to determining just where E is located, i.e., to determining the length of CE. Fermat's ultimate method for doing this (developed after his "double-root" approach) was to take O as an arbitrary point (between B and E) on the tangent line and then dropping the perpendicular from O to the x-axis (to I). Point A is the intersection of this perpendicular with the parabola. From the equation of that curve (remember, D is the origin) we have the distance relationships

$$CD = (BC)^2$$
$$ID = (AI)^2,$$

and so

$$\frac{(BC)^2}{(AI)^2} = \frac{CD}{ID}.$$

Since $OI > AI$, then

$$\frac{(BC)^2}{(OI)^2} < \frac{CD}{ID}.$$

   By similar triangles we also have

$$\frac{BC}{CE} = \frac{OI}{IE},$$

or

$$\frac{(BC)^2}{(OI)^2} = \frac{(CE)^2}{(IE)^2}.$$

Thus,

$$\frac{(CE)^2}{(IE)^2} < \frac{CD}{ID}.$$

Now, let $CD = d$, $CE = a$, and $CI = e$. Since B (and so C) is given, Fermat knew the value of d. The value of a is what Fermat wanted to calculate, while the value of e is variable as it depends on the choice for O (which determines I and so CI). In any case, we have $ID = CD - CI = d - e$, and $IE = CE - CI = a - e$, and so

$$\frac{a^2}{(a-e)^2} < \frac{d}{d-e},$$

or

$$a^2(d-e) < d(a-e)^2.$$

With a little algebra this becomes

$$2ade < a^2e + de^2.$$

To complete his argument, Fermat let O move ever closer to B, and this would of course move I ever closer to C, and so $e \to 0$. *But*, before doing that, $e \neq 0$, and so we can divide by e to get

$$2ad < a^2 + de.$$

Then letting $e \to 0$ transforms the inequality into an equality (obvious from the geometry of figure 4.4) and so $2ad = a^2$, or $d(= CD) = \frac{1}{2}a(= \frac{1}{2}CE)$. This is, of course, the same result he obtained from the double-root method, but *this* is the technique that so inspired Newton in his development of the differential calculus.

It didn't inspire Descartes, however, who thought the approach not to be general. He believed it would work only if an explicit relation of the form $y = y(x)$ could be written. In what he thought would convince Fermat (and others) that Fermat's method wouldn't be able to handle a curve more complicated than a mere parabola, Descartes challenged (in 1638) Fermat to apply it to the curve $x^3 +$

$y^3 = 3axy$, where a is a given positive constant. Notice that the x and y cannot be separated in this equation into the form $y = y(x)$. It is amusing to learn that, in addition to Descartes' failure to correctly draw his own curve, Fermat *was* able to quickly determine the tangent to the curve (now known as the "folium of Descartes.")

4.5 Derivatives and Tangents

The intimate connection between the derivative of a function $f(x)$ and the tangent to the curve $y = f(x)$ was used by Newton to solve the practical problem of calculating the roots to the equation $f(x) = 0$. As is now well known, if $f(x)$ is a polynomial of degree greater than four then there is no analytic solution, in general. What is called "Newton's method" is an iterative, numerical technique (see the next box) that can find the solutions to $f(x) = 0$ quickly, to any degree of accuracy desired, even in cases where $f(x)$ is a polynomial of *infinite* degree, e.g., $f(x) = x - \cos(x)$. Newton wrote up his discovery in 1671, as part of his book *Methodus fluxionum et serierum infinitarum*, but it wasn't actually published until 1736. Meanwhile, in 1690, the English mathematician Joseph Raphson (1648–1715) published the same method in his *Analysis aequationum universalis*. In modern calculus textbooks, this method (easily programmed on a computer) is often called the Newton-Raphson method in honor of both men.

To understand the geometry behind the Newton-Raphson method, let's consider the continuous function $f(x) = x^3 - 2x - 5$, the same function used by Newton in his *Method of Fluxions* to illustrate the method. It is easy to calculate that $f(2)$ and $f(3)$ have opposite algebraic signs, and so there must be some value of $x = \hat{x}$ (between 2 and 3) where $f(\hat{x}) = 0$. Figure 4.7, which plots $f(x)$, shows that the value of $\hat{x}$ is actually between 2 and 2.5, but suppose we want to find $\hat{x}$ much more precisely—e.g., accurate let's say, to *ten* decimal places? How can we do *that*?

The Newton-Raphson method generates a *sequence* of values, x_n for $n = 1, 2, 3, \cdots$, that approach $\hat{x}$, i.e., $\lim_{n \to \infty} x_n = \hat{x}$. That is, given the value x_k, the method then calculates x_{k+1} that is *closer* to $\hat{x}$; $|x_{k+1} - \hat{x}| < |x_k - \hat{x}|$. The Newton-Raphson method can then use x_{k+1} to calculate x_{k+2}, and so on, until we have the accuracy we desire. Here's how it works.

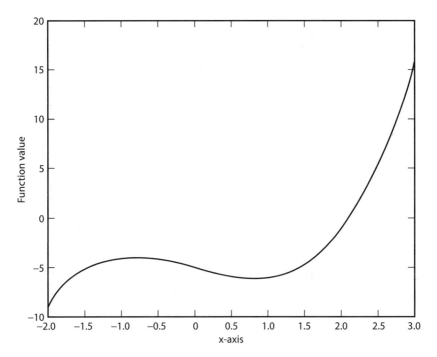

FIGURE 4.7. Newton's function.

The derivative of $f(x)$ at $x = x_n$ is $f'(x_n)$, which is the slope of the line tangent to $y = f(x)$ at $x = x_n$. This tangent line thus has the equation

$$y = f'(x_n)x + b,$$

where b is a constant. But, since $y = f(x_n)$ at $x = x_n$, then

$$f(x_n) = f'(x_n)x_n + b,$$

and so $b = f(x_n) - f'(x_n)x_n$. Thus, the tangent line has the equation

$$y = f'(x_n)x + f(x_n) - f'(x_n)x_n.$$

This tangent line crosses the x-axis (and so $y = 0$) at $x = x_{n+1}$, our next (often better, although not always—see figures 4.8a and 4.8b) approximation to $\hat{x}$. Thus,

$$0 = f'(x_n)x_{n+1} + f(x_n) - f'(x_n)x_n,$$

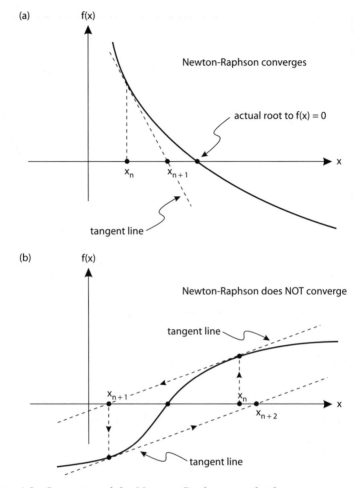

FIGURE 4.8. Geometry of the Newton-Raphson method.

or, solving for x_{n+1}, we have our result:

$$x_{n+1} = x_n - \frac{f(x_n)}{f'(x_n)}.$$

For Newton's example, $f'(x) = 3x^2 - 2$ and so the iterative algorithm for solving $f(x) = 0$ is

$$x_{n+1} = x_n - \frac{x_n^3 - 2x_n - 5}{3x_n^2 - 2} = \frac{2x_n^3 + 5}{3x_n^2 - 2}.$$

All we need now, to use this algorithm, is a "starting value" for the sequence of x_n, i.e., the value of x_0, the obvious choice for which is 2. Subsequent values generated by the algorithm are

$$x_1 = 2.1$$
$$x_2 = 2.09456812110419$$
$$x_3 = 2.09455148169820$$
$$x_4 = 2.09455148154233$$
$$x_5 = 2.09455148154233,$$

and so, after just four iterations we have the value of $\hat{x}$ to *better* than ten decimal places. The Newton-Raphson method itself is nothing but arithmetic, but it is fundamentally based on the connection between the derivative of a function and the tangent line (at a given point) to the curve determined by that function.

Relatively recent scholarship, I should tell you, convincingly argues that neither Newton *or* Raphson should have *this* method named after them! The method I just illustrated is both iterative and employs the derivative concept. Newton's own, specific calculation of the solution to the cubic has neither feature, and Raphson's method does not use derivatives (although it *is* iterative). It was actually the English mathematician Thomas Simpson (1710–61) who published the modern algorithm in 1740. For more on this interesting story, which has not yet (as far as I know) been incorporated into modern textbooks on the history of mathematics, see Nick Kollerstrom, "Thomas Simpson and 'Newton's Method of Approximation': An enduring myth" (*The British Journal for the History of Science*, September 1992, pp. 347–54).

Fermat came as close as one could to discovering the derivative *without* actually making the discovery. An "infinitesimal miss," yes, but for the credit of being declared the inventor of the differential calculus it made all the difference in the world. He *could* have taken the final step, too, as I'll illustrate in the next section on how Fermat finally constructed a proper derivation of Snell's law.

As one last example of the connection between derivatives and tangents, consider the problem of calculating the derivative of the function $f(x) = \ln(x)$. Using Fermat's idea, let's write

$$\frac{df}{dx} = \lim_{\Delta x \to 0} \frac{\ln(x + \Delta x) - \ln(x)}{\Delta x} = \lim_{\Delta x \to 0} \frac{\ln\left(\frac{x + \Delta x}{x}\right)}{\Delta x}$$

$$= \lim_{\Delta x \to 0} \frac{1}{\Delta x} \cdot \ln\left(1 + \frac{\Delta x}{x}\right) = \lim_{\Delta x \to 0} \ln\left(1 + \frac{\Delta x}{x}\right)^{1/\Delta x}.$$

Recall now that $\lim_{s \to \infty}(1+(a/s))^s = e^a$. If you don't recall this, there is a nice *noncalculus* derivation of it, using the binomial theorem, in Eli Maor's *e: The Story of a Number* (Princeton University Press, 1994, p. 35.) So, with $s = 1/\Delta x$, and $a = 1/x$, we have

$$\frac{d}{dx}\ln(x) = \lim_{\Delta x \to 0} \ln\left(1 + \frac{\Delta x}{x}\right)^{1/\Delta x} = \lim_{s \to \infty} \ln\left(1 + \frac{1/x}{s}\right)^s$$

$$= \ln\left(e^{1/x}\right) = \frac{1}{x}.$$

Figure 4.9 shows plots of $\ln(x)$ and $1/x$, and it is immediately obvious that $1/x$ does indeed "look like" the slope of $\ln(x)$. I'll use this result in the opening section of the next chapter to answer a famous "puzzle problem" in mathematics.

One of the most valuable of the differentiation rules tells us how to differentiate what are called *composite* functions. For example, we just learned what the derivative of $\ln(x)$ is, but what is the derivative of $\ln\{v(x)\}$, where $v(x)$ is *any* function of x, not simply $v(x) = x$? What, for example, is the derivative of $\ln\{\ln(x)\}$? The very definition of the derivative is the key to answering this. So, suppose $u = u(x)$ and $v = v(x)$, and that we already know how to differentiate $u(x)$ and $v(x)$, individually. We can find the derivative of $u\{v(x)\}$ by first writing

$$\frac{du}{dv} = \lim_{\Delta v \to 0} \frac{u(v + \Delta v) - u(v)}{\Delta v} \quad \text{and} \quad \frac{dv}{dx} = \lim_{\Delta x \to 0} \frac{v(x + \Delta x) - v(x)}{\Delta x}.$$

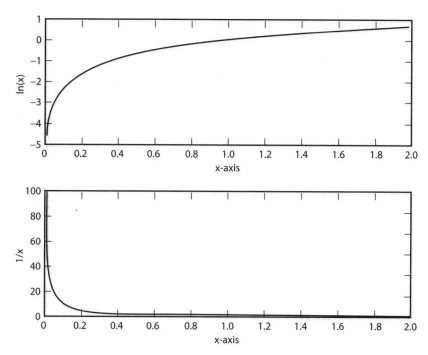

FIGURE 4.9. The natural log function and its derivative.

Also,

$$\frac{d}{dx}u\{v(x)\} = \lim_{\Delta x \to 0} \frac{u\{v(x + \Delta x)\} - u\{v(x)\}}{\Delta x}$$

$$= \lim_{\Delta x \to 0} \frac{u\{v(x + \Delta x)\} - u\{v(x)\}}{v(x + \Delta x) - v(x)} \cdot \frac{v(x + \Delta x) - v(x)}{\Delta x}.$$

Now, by definition $\Delta v = v(x + \Delta x) - v(x)$, and so $v(x + \Delta x) = v(x) + \Delta v$, which means that

$$\frac{d}{dx}u\{v(x)\} = \lim_{\Delta x \to 0} \frac{u\{v(x) + \Delta v\} - u\{v(x)\}}{\Delta v} \cdot \frac{v(x + \Delta x) - v(x)}{\Delta x}.$$

Since $\Delta v \to 0$ as $\Delta x \to 0$, we thus have

$$\frac{d}{dx}u\{v(x)\} = \lim_{\Delta v \to 0} \frac{u(v + \Delta v) - u(v)}{\Delta v} \cdot \lim_{\Delta x \to 0} \frac{v(x + \Delta x) - v(x)}{\Delta x}$$

$$= \frac{du}{dv} \cdot \frac{dv}{dx},$$

a result commonly called the *chain rule* and known to Leibniz no later than 1676. I used it in chapter 1 (in the minimum escape velocity problem of section 1.6), and I'll use it in the next chapter to solve a famous problem from 1686.

Now, to answer our original question on how to differentiate $\ln\{v(x)\}$, we have $u = \ln(v)$ and $v(x) = \ln(x)$, and so

$$\frac{d}{dx}\ln\{v(x)\} = \frac{d}{dv}\ln(v) \cdot \frac{dv}{dx} = \frac{1}{v}\frac{dv}{dx}.$$

For example, if $v(x) = \ln(x)$, then we have

$$\frac{d}{dx}\ln\{\ln(x)\} = \frac{1}{\ln(x)} \cdot \frac{1}{x} = \frac{1}{x\ln(x)},$$

which of course is defined only for $x > 1$.

And finally, we can turn all of this on its head and calculate the derivative of $f(x) = e^x$. This means $x = \ln f(x)$, and so

$$\frac{d}{dx}x = 1 = \frac{d}{dx}\ln\{f(x)\}.$$

But, our result for composite functions says

$$\frac{d}{dx}\ln\{f(x)\} = \left\{\frac{d}{df}\ln(f)\right\} \cdot \left\{\frac{df}{dx}\right\} = \frac{1}{f}\frac{df}{dx}.$$

So,

$$1 = \frac{1}{f}\frac{df}{dx},$$

or

$$\frac{df}{dx} = f, \quad \text{i.e.,} \quad \frac{d}{dx}e^x = e^x.$$

The exponential function is its own derivative.

Two highly useful results that immediately follow from this unique property of the exponential are the derivatives of the hyperbolic functions. Thus, with A some constant, if

$$f(x) = \cosh(Ax) = \frac{e^{Ax} + e^{-Ax}}{2}$$

$$g(x) = \sinh(Ax) = \frac{e^{Ax} - e^{-Ax}}{2},$$

then

$$\frac{df}{dx} = \frac{d}{dx}\cosh(Ax) = \frac{Ae^{Ax} - Ae^{-Ax}}{2} = A\,\sinh(Ax)$$

$$\frac{dg}{dx} = \frac{d}{dx}\sinh(Ax) = \frac{Ae^{Ax} + Ae^{-Ax}}{2} = A\,\cosh(Ax).$$

These formulas will be very helpful in chapter 6.

4.6   Snell's Law and the Principle of Least Time

Fermat's solution to finding a physically correct derivation of Snell's law of refraction was the result of developing a generalization of Heron's derivation of the reflection law. Using Heron's original minimum-path-length criterion wouldn't work for refraction, of course, as *that* path would simply be the straight line connecting A and B (in figure 4.10, where A and B have a lateral separation of d), rather than the actual broken path ARB. Fermat's generalization was to argue that the correct path, for both reflection *and* refraction, is the path of minimum *time*. For reflection, where the light is always in the same medium, minimum length and minimum time give the same path. But for refraction, the paths are different, and the least-time path is indeed the actual path. In the notation of figure 4.10, then, the total transit time from A to B is

$$T = \frac{\sqrt{h_1^2 + x^2}}{v_1} + \frac{\sqrt{h_2^2 + (d - x)^2}}{v_2}.$$

The mathematical problem for Fermat was to determine $x = \hat{x}$ so that T is minimized. One of the reasons why Fermat is not recognized as the inventor of the differential calculus is that he failed to

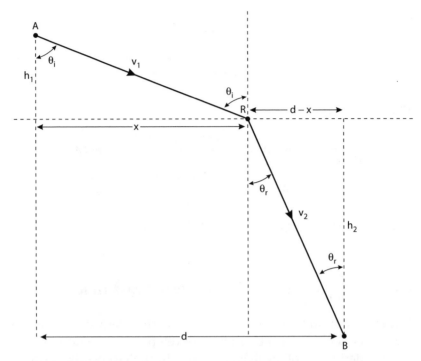

FIGURE 4.10. Geometry of Snell's law from the principle of least time.

discover the rules for applying his basic idea of $f(\hat{x} + E) \approx f(\hat{x})$ for E "small" to functions more complicated than simple polynomials, e.g., to the square roots that appear in the formula for T. Fermat was, however, through some special algebraic manipulations, still able to solve this specific problem. Here's how.

Let's start by observing from figure 4.10 that

$$\sin(\theta_i) = \frac{x}{\sqrt{h_1^2 + x^2}}$$

$$\sin(\theta_r) = \frac{d - x}{\sqrt{h_2^2 + (d - x)^2}}.$$

Then, using $T(\hat{x}) - T(\hat{x} + E) \approx 0$ for E "nearly zero," we have

$$\left[\frac{\sqrt{h_1^2 + \hat{x}^2}}{v_1} + \frac{\sqrt{h_2^2 + (d - \hat{x})^2}}{v_2} \right]$$

$$- \left[\frac{\sqrt{h_1^2 + (\hat{x} + E)^2}}{v_1} + \frac{\sqrt{h_2^2 + (d - \hat{x} - E)^2}}{v_2} \right] \approx 0.$$

Look now at the square root in the first term in the second pair of brackets; we can write it as (since E is "nearly zero" then E^2 is even "more nearly zero")

$$\sqrt{h_1^2 + (\hat{x} + E)^2} = \sqrt{h_1^2 + \hat{x}^2 + 2\hat{x}E + E^2} \approx \sqrt{h_1^2 + \hat{x}^2 + 2\hat{x}E}$$

$$= \sqrt{(h_1^2 + \hat{x}^2)\left(1 + \frac{2\hat{x}E}{h_1^2 + \hat{x}^2}\right)}$$

$$= \sqrt{1 + \frac{2\hat{x}E}{h_1^2 + \hat{x}^2}} \sqrt{h_1^2 + \hat{x}^2}.$$

Recalling once again the approximation $\sqrt{1 + u} \approx 1 + \frac{1}{2}u$ for u "small," we arrive at

$$\sqrt{h_1^2 + (\hat{x} + E)^2} \approx \left[1 + \frac{\hat{x}E}{h_1^2 + \hat{x}^2}\right] \sqrt{h_1^2 + \hat{x}^2}.$$

Repeating this process for the second square root (and using $\sqrt{1 - u} \approx 1 - \frac{1}{2}u$) results in

$$\sqrt{h_2^2 + (d - \hat{x} - E)^2} \approx \left[1 - \frac{(d - \hat{x})E}{(d - \hat{x})^2 + h_2^2}\right] \sqrt{h_2^2 + (d - \hat{x})^2}.$$

We thus have

$$T(\hat{x}) - T(\hat{x} + E) = \left[\frac{\sqrt{h_1^2 + \hat{x}^2}}{v_1} - \frac{\sqrt{h_1^2 + (\hat{x} + E)^2}}{v_1} \right]$$

$$+ \left[\frac{\sqrt{h_2^2 + (d - \hat{x})^2}}{v_2} - \frac{\sqrt{h_2^2 + (d - \hat{x} - E)^2}}{v_2} \right]$$

$$\approx \frac{\sqrt{h_1^2 + \hat{x}^2}}{v_1} \left[1 - \left\{ 1 + \frac{\hat{x}E}{h_1^2 + \hat{x}^2} \right\} \right]$$

$$+ \frac{\sqrt{h_2^2 + (d - \hat{x})^2}}{v_2} \left[1 - \left\{ 1 - \frac{(d - \hat{x})E}{(d - \hat{x})^2 + h_2^2} \right\} \right]$$

$$= \frac{\sqrt{h_2^2 + (d - \hat{x})^2}}{v_2} \cdot \frac{(d - \hat{x})E}{(d - \hat{x})^2 + h_2^2} - \frac{\sqrt{h_1^2 + \hat{x}^2}}{v_1} \cdot \frac{\hat{x}E}{h_1^2 + \hat{x}^2} \approx 0.$$

Next, dividing through by E (which is not yet *exactly* zero) and *then* imagining E vanishes, we arrive at the equality

$$\frac{\sqrt{h_2^2 + (d - \hat{x})^2}}{v_2} \cdot \frac{(d - \hat{x})}{(d - \hat{x})^2 + h_2^2} = \frac{\sqrt{h_1^2 + \hat{x}^2}}{v_1} \cdot \frac{\hat{x}}{h_1^2 + \hat{x}^2},$$

or

$$\frac{1}{v_2} \cdot \frac{d - \hat{x}}{\sqrt{h_2^2 + (d - \hat{x})^2}} = \frac{1}{v_1} \cdot \frac{\hat{x}}{\sqrt{h_1^2 + \hat{x}^2}}.$$

But this is just

$$\frac{1}{v_2} \sin(\theta_r) = \frac{1}{v_1} \sin(\theta_i),$$

or, at last,

$$\frac{\sin(\theta_i)}{\sin(\theta_r)} = \frac{v_1}{v_2} = \text{constant}.$$

I say *at last* because while this is once again our now familiar Snell's law, *now* we (Fermat) have the constant right! It is v_1/v_2, the inverse of Descartes' result of v_2/v_1, which had forced him to conclude that v_2 (= the speed of light in water) $>$ v_1 (= the speed of light in air) because experiment shows the constant in Snell's law is greater than one. For Fermat, however, the conclusion was just the reverse: $v_2 < v_1$.

Fermat was both astonished and pleased at this success of his principle of least time, as it simultaneously explained how Descartes' result could be in agreement with experiment and at the same time *wrong* in its conclusion about the speed of light in different mediums. A modern student would, of course, be perplexed at all of the *algebra* Fermat used. She would wonder at why he hadn't simply set the derivative of T equal to zero to find Snell's law. The answer is, as I mentioned earlier, that Fermat didn't know how to do that. But he *was* so very close.

Indeed, the general differentiation formulas for Fermat's problem are not hard to develop and, in June 1682, Leibniz carried out the following analysis. Looking at the expression for T, we see that we have just two fundamental forms: if c_1 and c_2 are constants, the forms are

$$g(x) = c_1 + (c_2 - x)^2$$

$$h(x) = \sqrt{c + g(x)}.$$

In section 4.4 you saw how to differentiate $h(x)$. To differentiate $g(x)$, we write (using Fermat's basic idea) in modern notation,

$$\frac{dg}{dx} = \lim_{\varepsilon \to 0} \frac{g(x+\varepsilon) - g(x)}{\varepsilon}.$$

So,

$$\frac{dg}{dx} = \lim_{\varepsilon \to 0} \frac{\left[c_1 + \{c_2 - (x+\varepsilon)\}^2\right] - \left[c_1 + (c_2 - x)^2\right]}{\varepsilon}$$

$$= \lim_{\varepsilon \to 0} \frac{\left[c_1 + c_2^2 - 2c_2(x+\varepsilon) + (x+\varepsilon)^2\right] - \left[c_1 + c_2^2 - 2c_2 x + x^2\right]}{\varepsilon}$$

$$= \lim_{\varepsilon \to 0} \frac{-2c_2(x+\varepsilon) + (x+\varepsilon)^2 + 2c_2 x - x^2}{\varepsilon}$$

$$= \lim_{\varepsilon \to 0} \frac{-2c_2x - 2c_2\varepsilon + x^2 + 2x\varepsilon + \varepsilon^2 + 2c_2x - x^2}{\varepsilon}$$

$$= \lim_{\varepsilon \to 0} \frac{-2c_2\varepsilon + 2x\varepsilon + \varepsilon^2}{\varepsilon} = \lim_{\varepsilon \to 0} (-2c_2 + 2x + \varepsilon)$$

$$= -2(c_2 - x).$$

And from section 4.4 we have

$$\frac{dh}{dx} = \frac{1}{2} \cdot \frac{1}{\sqrt{c + g(x)}} \cdot \frac{dg}{dx}.$$

With these formulas in hand, the modern student would take T, written as

$$T = \frac{\sqrt{h_1^2 + x^2}}{v_1} + \frac{\sqrt{h_2^2 + (d - x)^2}}{v_2},$$

and, as did Leibniz, write (*by inspection*)

$$\frac{dT}{dx} = \frac{1}{2} \cdot \frac{1}{v_1} \cdot \frac{1}{\sqrt{h_1^2 + x^2}}[2x] + \frac{1}{2} \cdot \frac{1}{v_2} \cdot \frac{1}{\sqrt{h_2^2 + (d-x)^2}}[-2(d-x)] = 0,$$

or

$$\frac{x}{v_1\sqrt{h_1^2 + x^2}} - \frac{d - x}{v_2\sqrt{h_2^2 + (d - x)^2}} = 0.$$

Recalling the expressions for $\sin(\theta_i)$ and $\sin(\theta_r)$, this immediately reduces to Snell's law,

$$\frac{\sin(\theta_i)}{\sin(\theta_r)} = \frac{v_1}{v_2}.$$

Fermat's principle of least time does strike many as being outside of mathematics, and perhaps even outside of physics as well; as being *metaphysical*. Of course, Heron's derivation of the law of reflection from the principle of minimum distance is open to the same criticism (and, obviously, minimum distance is equivalent to minimum time for travel always in the same medium, and so Heron's principle is simply a special case of Fermat's). Students always want to know *how* does light "know," at the start of a journey, what path will result in minimum length (time)? That seems to require light to be prescient! Before the development of quantum electrodynamics, which explains how light "knows," Fermat's principle *did* have to be taken on faith, and for many that was too much to ask. Fermat himself was not sympathetic to those who rejected the least-time principle on the grounds that it asked for light to know where it was going before it started. As he replied in a (unconscious?) pun to one of his critics, "I do not pretend to be in the secret confidence of nature. She works by paths obscure and hidden. . . ."

In fact, Fermat's principle of *least* time is *not* always correct. The modern statement of the principle says the path a light beam follows is simply a *stationary* path, which means that a *slight variation* in the optical path leaves the travel time unchanged. This may indeed result in a path with minimum travel time, but another possibility is a path with *maximum* travel time! To see how such a thing could happen, imagine a point source of light in the center (point O) of an ellipsoidal mirror, as shown in figure 4.11. There are four points around the mirror (A, B, C, and D) which reflect light directly back to O. Two of them (A and B) determine minimum time paths, while the other two (C and D) determine maximum time paths.

The criticism Fermat received about the principle of least time was slight indeed compared to that which descended upon Pierre Louis Moreau de Maupertuis (1698–1759) over his so-called principle of least action. A number of people, long before Fermat (and probably

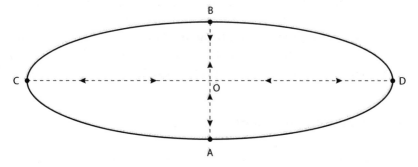

FIGURE 4.11. Minimum and maximum time paths are stationary paths.

even before Heron) had thought that a universe made by God must be a perfect universe, and consequently should always operate with economy. Leonardo da Vinci, for example, who wasn't really even a very good mathematician, nevertheless was a thoughtful intellect and declared (more than a century before Fermat) that "Every action done by nature is done in the shortest way." He failed, however, to explain just what that might mean. Fermat added an explanation for the case of light, but Maupertuis went light-years further by both defining *action* and claiming "least action" to be universally applicable: "in all the changes that take place in the universe, the sum of the products of each body multiplied by the distance it moves and by the speed with which it moves is the least possible." He published this in 1746, shortly after becoming President of the Academy of Sciences in Berlin. Least action was later made more precise by such giants as Euler, Hamilton, and Lagrange, and it has found enormously fruitful applications in such diverse fields as physics (quantum mechanics) and biology (self-regulating, living systems).

For Maupertuis, however, who seemingly was guided more by theological reasoning than by mathematical physics, least action brought ridicule down on his head, with the worst of it coming from his one-time friend Voltaire. That argument over least action became one of the nastiest scientific brawls in history, and it was initiated by a claim from a mathematician named König (for more on him, see the end of appendix C) that, first, it was wrong, and second, that Maupertuis had stolen it anyway from an unpublished 1707 letter by Leibniz! Euler declared Maupertuis was right, but he was no match for the poison-pen of math-illiterate Voltaire; both Euler and Maupertuis were the initial losers in this battle. Today we better understand who was right and who was not, but that is of little consequence for the dead. You can read more about this savage, bitter controversy in the essay by Bently Glass, "Maupertuis, Pioneer of Genetics and Evolution," included in *Forerunners of Darwin, 1745–1859* (The Johns Hopkins University Press 1959).

4.7 A Popular Textbook Problem

The refraction of light and Fermat's principle of least time have served as the inspiration for numerous calculus textbook problems

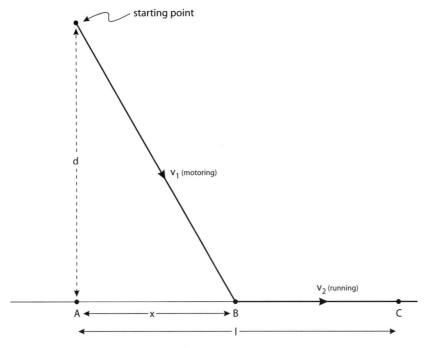

FIGURE 4.12. Geometry of yet another minimum-time lake-crossing problem.

of the following type (illustrated in figure 4.12). A man is in a power-boat in a lake, distance d from the nearest point (A) on the shore (which is taken to be straight). He wishes to travel, by a combination of motoring and running, to point C on the shore. Point C is a distance ℓ from A. That is, he will motor directly to some point B on the shore, distance x from A, and then run from B to C. If the boat travels at speed v_1 and if the man runs at speed v_2, then what is x (where is B?) so as to minimize his total travel time? To be as general as possible, we'll consider both the case of $v_1 < v_2$ and of $v_1 > v_2$. This problem, even though less sophisticated than the superficially similar problem at the end of chapter 1, is worth some attention here because it has an easy-to-miss, subtle issue.

We start by writing the total travel time T, as a function of x, as

$$T(x) = \frac{\sqrt{d^2 + x^2}}{v_1} + \frac{\ell - x}{v_2}.$$

Following the standard prescription for finding an extrema (a minimum for T) we set the derivative of $T(x)$ to zero:

$$\frac{dT}{dx} = \frac{1}{2} \cdot \frac{1}{\sqrt{d^2 + x^2}} \cdot 2x \cdot \frac{1}{v_1} - \frac{1}{v_2} = 0.$$

With just a little algebra this is easily solved to give

$$x = \frac{\left(\dfrac{v_1}{v_2}\right) d}{\sqrt{1 - \left(\dfrac{v_1}{v_2}\right)^2}}.$$

It does seem a bit odd that this formal result for x (the location of B) is independent of ℓ, and so we might well wonder if this formal result is actually correct. Well, it *might* be correct—but not necessarily! Here's why.

If $(v_1/v_2) > 1$ (if the boat travels faster than the man runs), then there is no *real* formal solution for x_1 because the denominator is imaginary. The physical interpretation for this case is simply that $x = \ell$, i.e., the man should motor straight, all the way, to C. This is, of course, the obvious statement that if the boat moves faster than the man can run, then the shortest total travel time is achieved by always traveling at the greater speed along the shortest path (the straight line segment joining his initial position directly with C).

But even if $v_1/v_2 < 1$ (and so the formal solution for x is real) it is not always the *correct* solution. This is because it is physically obvious that, for *any* values of v_1 and v_2, x will be confined to the interval $0 \le x \le \ell$. After all, it makes no sense to motor to either an $x > \ell$ or an $x < 0$ and then run all the way back to C! Now, x is confined to this interval only if $0 \le v_1/v_2 \le m$, where m is the finite value of v_1/v_2 that gives $x = \ell$ (the condition $v_1/v_2 = 0$ gives the other extreme, so-called *end-point* value for x, that is, $x = 0$). We can solve for m by setting

$$\frac{\left(\dfrac{v_1}{v_2}\right) d}{\sqrt{1 - \left(\dfrac{v_1}{v_2}\right)^2}} = \ell,$$

which gives

$$\left(\frac{v_1}{v_2}\right)_{\max} = m = \frac{1}{\sqrt{1 + \left(\frac{d}{\ell}\right)^2}}.$$

So, even if $v_1 < v_2$, there is the possibility that the straight path from A to C is the minimum-time path.

The answer to this problem is therefore actually *not* independent of ℓ, as the formal result misleadingly suggests. That is,

$$if\ 0 \le \frac{v_1}{v_2} \le \frac{1}{\sqrt{1 + \left(\frac{d}{\ell}\right)^2}} \quad then \quad x = \frac{\left(\frac{v_1}{v_2}\right)d}{\sqrt{1 - \left(\frac{v_1}{v_2}\right)^2}},$$

$$otherwise\ x = \ell.$$

The moral is obvious: the solution to a minimization problem *may* be given by the vanishing of a derivative, but then it may also *not* be! This important conclusion is forgotten at the analyst's peril.

4.8 Snell's Law and the Rainbow

Physicists write Snell's law in a slightly different manner than we have so far used, with c denoting the speed of light in a vacuum:

$$\frac{\sin(\theta_i)}{\sin(\theta_r)} = \frac{v_1}{v_2} = \frac{c/v_2}{c/v_1} = \frac{n_2}{n_1},$$

where $n_1 = c/v_1$ and $n_2 = c/v_2$ are called the *indices of refraction* for medium 1 and medium 2, respectively. That is, the index of refraction for a medium is simply the ratio of the speed of light in a vacuum to the speed of light in the medium. The usual case is, of course, that the index of refraction is a positive number greater than 1. For the normal mediums of air, water, and glass, the indices of refraction are usually taken to be 1, 1.333, and 1.5, respectively, but these are really just typical values. The index of refraction for a given medium isn't just a single number, but rather is a function of the frequency (wavelength) of the light. For example, the index of refraction for water decreases with increasing wavelength; in the visible portion of the electromagnetic spectrum (the so-called *optical* region), as light varies through the colors violet ("short" wavelength), blue, green, yellow, orange, to red ("long" wavelength), the index of refraction varies from 1.344 to 1.331.

The fact that the index of refraction for a medium depends on the frequency of the light explains why what appears to be white light can be separated by refraction into various colored constituents. Each colored component of the total white light experiences a slightly different angle of refraction in Snell's law, and so is separated from its other differently colored (different wavelength) companions. This effect, called dispersion, was discovered by Newton in his famous glass prism experiment of 1666 (after both Descartes and Fermat were dead).

With Snell's law written as

$$\sin(\theta_r) = \frac{n_1}{n_2}\sin(\theta_i),$$

we can see that if $n_2 > n_1$ (as is the case when light, in air, is incident on a water surface), then $\theta_r < \theta_i$. That is, the refracted light is bent toward the normal. However, since θ_r is still positive, the refracted light is not bent *beyond* the normal. The bent light beam travels into the water on the opposite side of the normal from the incident light. The contrary case, never seen in nature, would mean $\theta_r < 0$ and thus require a *negative* index of refraction. But would such a thing be *impossible*?

In the late 1960s, theoretical studies in the Soviet Union showed that a negative index medium, while undeniably strange, would not violate any of the fundamental laws of physics. In 2001, American physicists at the University of California/San Diego actually fabricated what they call a "structured metamaterial" that, in the microwave frequency band of 10.2 to 10.6 GHz, has a negative index of refraction. This is an extremely high frequency by many standards, e.g., the middle of the AM radio frequency band is one megahertz = 0.001 GHz. Ten gigahertz, however, is a very low frequency compared to optical frequencies (on the order of 500,000 GHz), and whether or not negative index optical frequency devices can be made is still very much an open question. See R. A. Shelby, et. al, "Experimental Verification of a Negative Index of Refraction" *(Science*, April 6, 2001, pp 77–79).

My reason for getting into the physics of refraction as much as I have is that Descartes next used Snell's law to explain, using a maximum argument, the first mystery of the rainbow: *why* there is often a bright, circular arc of light in the sunlit sky after a rainstorm. You'll see how he did this in the next chapter, and how calculus (which he did *not* use) is the perfect tool with which to study the rainbow. The second mystery of the rainbow (why is it a *multicolored* arc of light, and not just a white arc?) remained a mystery to Descartes because he didn't know about dispersion, and so he used a single number for the index of refraction for water (droplets in the sky). Descartes did have an "explanation" for the colors, but it is (like his "derivation" of Snell's law itself) physical nonsense. My second reason for discussing the physics of the refraction of light is that, in 1696, the Swiss mathematician Johann Bernoulli used Snell's law to solve a physics minimization problem (discussed in chapter 6) that marks the origin of the calculus of variations, the next step up in advanced mathematics beyond the calculus itself.

5.

Calculus Steps Forward, Center Stage

5.1 The Derivative: Controversy and Triumph

Starting with Fermat's near miss of the derivative, and the later work by Newton and Leibniz, and others, in developing general differentiation formulas, the differential and integral calculus had, by 1700, become *the* mathematics for solving many (but not all, as you'll see when we get to later chapters) extrema problems. But not *everybody* was convinced that a quantum leap in mathematics had been achieved. As late as 1734, for example, the British philosopher George Berkeley (1685–1753) could rightfully pen an attack on the logical foundations of calculus, as he did in *The Analyst: or a discourse addressed to an infidel mathematician*. His motivation for this was more theological than mathematical, however; appointed a bishop that same year, he wrote *The Analyst* as a rebuttal to those who were turning away from the faith and embracing instead the so-called rationality of mathematics and science. Bishop Berkeley thought that view misguided, writing in his polemic "He who can digest a second or third fluxion, . . . need not, we think, be squeamish about any point of divinity." Even more famous is his remark, also from *The Analyst*, which appears to try to tie calculus to the supernatural as much as to religion: "And what are the fluxions? The velocities of evanescent increments? They are neither finite quantities, nor quantities infinitely small, nor yet nothing. May we not call them ghosts of departed quantities?" Bishop Berkeley's hope of showing calculus to be fatally flawed failed in the long run, but his criticisms did result in mathematicians returning time and again to the vital task of placing calculus on a logically secure foundation.

Since the start of the eighteenth century calculus has leapt from one spectacular triumph to the next, and continues to this day to be the rite-of-passage from high school math to the so-called advanced maths. Calculus has earned this reputation because of its ability to successfully handle problems that, without it, are simply impossible. In this chapter I'll discuss a number of such problems, all mathematically interesting, with some also having important historical significance as well. So, to start, consider the following freshman calculus puzzle that has been known to drive even math professors to despair.

Imagine you are stranded on a desert island, with only a stick to write in the acres of sand that surround you. You certainly do *not* have a table of logarithms or a calculator! If asked "which is larger, 3^4 or 4^3?", you would have no problem scribbling the solution in the sand with your stick: $3^4 = 3 \cdot 3 \cdot 3 \cdot 3 = 81 > 4^3 = 4 \cdot 4 \cdot 4 = 64$. This is easy because 3 and 4 are (small) integers. But what if the question is "which is larger, e^π or π^e?"? Both e and π are transcendental, and that complicates matters (how do you write e π times, or π e times?). Since both e and π are close to 3 you would probably correctly guess that the two expressions have nearly the same value, but that doesn't tell us which is the larger. What to do? With the derivative, it is "easy" (it's *always* easy, if you think of the right approach).

Start by defining the function $h(x) = (\ln(x)/x)$ (thinking of this definition is the "hard" part of the problem!). With $f(x)$ and $g(x)$ as two functions of x, such that

$$h(x) = \frac{f(x)}{g(x)},$$

one of the fundamental differentiation formulas of calculus tells us that

$$\frac{d}{dx} h(x) = \frac{d}{dx}\left\{\frac{f(x)}{g(x)}\right\} = \frac{g(x)\dfrac{df}{dx} - f(x)\dfrac{dg}{dx}}{g^2(x)}.$$

For example, since $\tan(x) = \sin(x)/\cos(x)$, and as $(d/dx)\sin(x) = \cos(x)$ and $(d/dx)\cos(x) = -\sin(x)$, results easily established with the fundamental definition of the derivative, we then have

$$\frac{d}{dx}\tan(x) = \frac{\cos(x)\dfrac{d}{dx}\sin(x) - \sin(x)\dfrac{d}{dx}\cos(x)}{\cos^2(x)} = \frac{\cos^2(x) + \sin^2(x)}{\cos^2(x)}$$

$$= \frac{1}{\cos^2(x)}.$$

Now, with $f(x) = \ln(x)$ and $g(x) = x$, we have from results established in the last chapter that

$$\frac{dh}{dx} = \frac{x\dfrac{d}{dx}\ln(x) - \ln(x)}{x^2} = \frac{x \cdot \dfrac{1}{x} - \ln(x)}{x^2} = \frac{1 - \ln(x)}{x^2}.$$

Thus, the derivative vanishes (our condition for an extrema) when $1 - \ln(x) = 0$, i.e., when $x = e$. But what *kind* of extrema does $x = e$ give us? Is it a minimum or a maximum? We can argue geometrically that it is a maximum, as the plot of $h(x)$ in figure 5.1 shows.

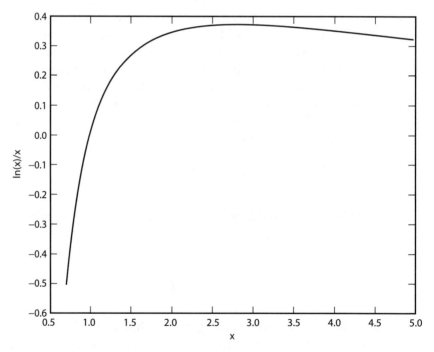

FIGURE 5.1. This function has a (broad) maximum at $x = e$ ($= 2.718\ldots$).

Geometrical arguments and plots are limited, however, to those situations where we can easily see "what is going on" with the function of interest. More generally, we need an analytical way to distinguish minimums from maximums, and such a way is provided by the *second* derivative. Isaac Newton was the first (1665) to see this and, ironically, the basic idea behind this *analytical* method is intuitively obvious if we look at it *physically*. So, to be specific, suppose $h(t)$ represents the height at time t of a ball thrown upward. Then dh/dt is the *speed* of the ball, and $dh/dt = 0$ simply says that the ball has an instantaneous speed of zero at its maximum height, i.e., it has stopped moving upward (positive speed) and is about to begin its fall back to the ground (negative speed because the direction of motion is reversed).

The second derivative, d^2h/dt^2, is the rate of change of the speed, i.e., it is the ball's *acceleration* (due entirely to the force of gravity). But that force is always pointed downward toward the center of the Earth, opposite to the direction of increasing $h(t)$. Thus, $d^2h/dt^2 < 0$, *always*. This gives us the so-called second derivative test for a (local) maximum. If $d^2h/dt^2 < 0$ when $dh/dt = 0$, then $h(t)$ has an extrema that is a (local) maximum. If $d^2h/dt^2 > 0$ when $dh/dt = 0$, however, then $h(t)$ has an extrema that is a (local) minimum.

Bishop Berkeley was, as mentioned earlier, greatly distressed over the logical basis of the *first* derivative; one can easily imagine his horror at the second derivative. Indeed, here are his own words from *The Analyst*: "But the velocities of the velocities, the second, third, fourth, and fifth velocities, &c., exceed, if I mistake not, all human understanding. The further the mind analyseth and pursueth these fugitive ideas the more it is lost and bewildered. . . ."

In the above discussion, t (time) is the independent variable, but that is of no special consequence. The second derivative test applies equally well to functions of *any* independent variable, e.g., to the $h(x)$ of our original problem. So, calculating the second derivative, we have

$$\frac{d^2h}{dx^2} = \frac{x^2\left(-\dfrac{1}{x}\right) - [1 - \ln(x)]2x}{x^4} = \frac{-x - 2x + 2x\ln(x)}{x^4}.$$

Thus,

$$\frac{d^2h}{dx^2}\bigg|_{x=e} = \frac{-3e + 2e\ln(e)}{e^4} = \frac{-3e + 2e}{e^4} = -\frac{1}{e^3} < 0.$$

That is, $x = e$ is the location of the *maximum* of $h(x)$, just as illustrated in figure 5.1.

By the very meaning of *maximum*, any value of $x \neq e$, such as $x = \pi$, will give a *smaller* value for $h(x)$. Thus, for the $h(x)$ of our original problem,

$$\frac{\ln(e)}{e} = \frac{1}{e} > \frac{\ln(\pi)}{\pi},$$

or

$$\pi > e\ln(\pi) = \ln(\pi^e).$$

Thus,

$$e^\pi > e^{\ln(\pi^e)} = \pi^e,$$

and we are done. In fact, a calculator does confirm that $e^\pi = 23.14069$ . . . is indeed larger (but not by very much) than $\pi^e = 22.45915. . . .$

The differentiation rule for a quotient also quickly gives us the rule for differentiating a product, i.e., the formula for

$$\frac{d}{dx}\{f(x)\,u(x)\} = ?$$

If we define $u(x) = 1/g(x)$, then we have from before that,

$$\frac{d}{dx}\left\{\frac{f(x)}{g(x)}\right\} = \frac{g\dfrac{df}{dx} - f\dfrac{dg}{dx}}{g^2} = \frac{\dfrac{1}{u}\dfrac{df}{dx} - f\dfrac{d}{dx}\left(\dfrac{1}{u}\right)}{1/u^2}.$$

Applying the quotient rule to $(d/dx)[(1/u)]$, and remembering that the derivative of a constant is zero, we have

$$\frac{d}{dx}\left(\frac{1}{u}\right) = \frac{-\dfrac{du}{dx}}{1/u^2}.$$

and so

$$\frac{d}{dx}\{f(x)\,u(x)\} = \frac{\dfrac{1}{u}\dfrac{df}{dx} - f\left[\dfrac{-\dfrac{du}{dx}}{u^2}\right]}{1/u^2}$$

$$= u\,\frac{df}{dx} + f\,\frac{du}{dx}.$$

By 1677 the rules for differentiating quotients and products were known to Leibniz.

The rule for differentiating a product leads immediately to one of the fundamental results of *integral* calculus: the formula for integration-by-parts. If we take advantage of Leibniz's differential notation and "multiply through" by dx, then we obtain

$$d(fu) = u\,df + f\,du,$$

or

$$u\,df = d(fu) - f\,du.$$

Then, integrating from $x = a$ to $x = b$, we arrive at

$$\int_a^b u(x)\,df = \left\{ f(x)\,u(x) \,\Big|_a^b - \int_a^b f(x)du \right.$$

We'll use this result at a crucial point in chapter 6 when we derive the Euler-Lagrange differential equation, which is at the core of the calculus of variations.

What does it mean if, when $h'(x) = dh/dx = 0$, we have $h''(x) = d^2h/dx^2 = 0$ as well? The second derivative test, which asks if $h''(x)$ is either *greater* than or *less* than zero, would seem to be equivocating when $h''(x)$ is equal to zero. And indeed it is. In this case $h(x)$ may or may *not* have an extrema. It is easy to demonstrate both possibilities. Suppose $h(x) = x^3$. Then $h'(x) = 3x^2$ and $h''(x) = 6x$. Both derivatives vanish at $x = 0$, where there is *not* an extrema, as

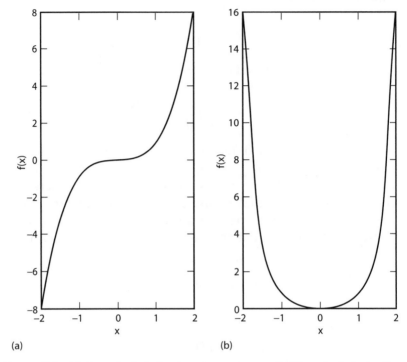

(a)   (b)

FIGURE 5.2. (a) Zero 2nd derivative, no extrema. (b) Zero 2nd derivative, with extrema.

shown in the first half of figure 5.2. However, if $h(x) = x^4$, then $h'(x) = 4x^3$ and $h''(x) = 12x^2$, and again both derivatives vanish at $x = 0$ where there *is* an extrema (a minimum), as shown in the second half of figure 5.2. We can distinguish the "extrema" and "no extrema" cases by observing that, for an extrema, $h''(x)$ does not change sign around the extrema (indeed, $12x^2$ *never* changes its sign), while when there is no extrema, $h''(x)$ does change its sign around the value of x that gives $h''(x) = 0$ ($6x$ does change sign around $x = 0$). In this last case, we say $h(x)$ has an *inflection point*.

Here's a pretty little differentiation problem of historical interest, using all of the above ideas, for you to try your hand at. Posed by the nineteenth-century Swiss mathematician Jacob

Steiner (mentioned in chapter 2 in connection with the isoperi-metric problem), it asks for the value of x for which the xth root of x is a maximum. That is, if we define $f(x)$ as

$$f(x) = \sqrt[x]{x} = x^{\frac{1}{x}}, \qquad x > 0,$$

then for what x is $f(x)$ the largest (and what is that maximum value)? Before starting your analysis you should convince your-self (with *non*calculus reasoning!) that, as x increases from zero, $f(x)$ first increases and then decreases, which suggests $f(x)$ does indeed have a maximum. The answer is at the end of this chapter.

5.2 Paintings Again, and Kepler's Wine Barrel

With the derivative, the original Regiomontanus problem from sec-tion 3.1, of determining the "best" distance to stand away from a painting hanging on a wall, becomes routine. In the notation of figure 3.2, the problem was to determine the x that maximizes θ in the expression

$$\tan(\theta) = \frac{(b-a)x}{x^2 + (b-h)(a-h)}.$$

Since $\tan(\theta)$ increases with increasing θ, then simply maximizing the right-hand side will also maximize θ. In chapter 3 we used a tricky, noncalculus approach. But now we can write

$$\frac{d}{dx}\tan(\theta) = \frac{d}{dx}\left\{\frac{f(x)}{g(x)}\right\} = 0,$$

with $f(x) = (b-a)x$ and $g(x) = x^2 + (b-h)(a-h)$, and solve. From the differentiation formula in the last section for a quotient, we see that this is equivalent to solving

$$g(x)\frac{df}{dx} = f(x)\frac{dg}{dx},$$

i.e., to solving

$$\left[x^2 + (b-h)(a-h)\right](b-a) = (b-a)x(2x).$$

This quickly results in $x = \sqrt{(b-h)(a-h)}$, just as we found in chapter 3.

Another historical problem that yields easily to the derivative is Kepler's wine barrel problem (mentioned in the previous chapter), on how to make the right cylindrical wine barrel of maximum volume and prescribed diagonal (ℓ). With the aid of the derivative, this is now a standard problem (in various disguises) in freshman calculus texts with, sadly, the history almost always unmentioned. In the notation of figure 5.3, where r, h, and V are the barrel's radius, height, and volume, respectively, we have

$$\ell^2 = (2r)^2 + h^2 = 4r^2 + h^2$$
$$V = \pi r^2 h.$$

So,

$$r^2 = \frac{\ell^2 - h^2}{4},$$

and thus

$$V = \pi \frac{\ell^2 - h^2}{4} h = \frac{\pi}{4}\left(\ell^2 h - h^3\right).$$

With V now expressed in terms of the single variable h, we can find the extrema of V by writing

$$\frac{dV}{dh} = 0 = \frac{\pi}{4}\left(\ell^2 - 3h^2\right),$$

which says $h^2 = \frac{1}{3}\ell^2$. Thus,

$$\ell^2 = (2r)^2 + \frac{1}{3}\ell^2,$$

or, with $d = 2r$ as the barrel's diameter, we have

$$\ell^2 = d^2 + \frac{1}{3}\ell^2,$$

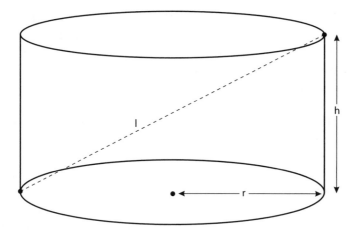

FIGURE 5.3. Kepler's wine barrel.

or, $d^2 = \frac{2}{3}\, \ell^2$. That is,

$$\frac{d^2}{h^2} = \frac{\dfrac{2}{3}\,\ell^2}{\dfrac{1}{3}\,\ell^2} = 2, \quad \text{i.e., } \frac{d}{h} = \sqrt{2},$$

as stated back in chapter 4. The actual volume of the largest barrel is

$$V_{\max} = \pi r^2 h = \pi \left(\frac{d}{2}\right)^2 h = \frac{\pi}{4}\, d^2 h = \frac{\pi}{4} \cdot \frac{2}{3}\, \ell^2 \cdot \frac{\ell}{\sqrt{3}}$$

$$= \frac{\pi}{6\sqrt{3}}\, \ell^3 = 0.3023\, \ell^3.$$

5.3   The Mailable Package Paradox

An interesting maximization problem of more recent vintage gives rise to the mailable package paradox. When you send a package by UPS (United Parcel Service), there are certain physical constraints you have to satisfy. These have changed over the years, but as I write, the maximum allowable length is 108", and the maximum *size*— defined by UPS as the length plus the package's maximum girth—is

130". The *girth* at any point along the length is the distance around the cross section at that point (taken perpendicular to the length). Since it is the maximum girth that is used to determine the size, then it is clear that to maximize the package's volume we should have *all* cross sections with the same girth. Since for a given girth (cross section perimeter) a *circular* cross section has the largest area, it then follows that a right circular cylinder is the shape of the maximum volume package (and *not* a sphere, as you'll soon see).

This seemingly peculiar definition of size (length plus maximum girth) is used instead of the more obvious one of volume because it is easier and faster for a mail agent to determine. All that is needed is a flexible measuring tape, and no complicated volume calculations are required (just addition). *But,* there is a price paid for the convenience of this definition: it does occasionally lead to a paradoxical result. That is, it is possible to make two packages that, when presented to a mail agent, are such that the agent will accept the larger volume but will reject the smaller volume! With calculus, this odd situation is easy to understand.

Consider a cylindrical package with maximum volume with radius r, length x, and volume v. If S denotes the maximum size allowed, then

$$v = \pi r^2 x \quad \text{and} \quad S = x + 2\pi r.$$

Thus,

$$r = \frac{S - x}{2\pi} \quad \text{and so} \quad v = \pi \left(\frac{S - x}{2\pi} \right)^2 x = \frac{1}{4\pi} \left(x S^2 - 2Sx^2 + x^3 \right).$$

We have an extrema for the volume when $dv/dx = 0$, i.e., when

$$S^2 - 4Sx + 3x^2 = 0.$$

This is easily solved to give either $x = S$ or $x = \frac{1}{3}S$. We reject the first solution because then $r = 0$, which certainly gives the *minimum* volume of zero! We therefore have $x = \frac{1}{3}S$ for the maximum volume (I'll leave it for you to verify that $d^2v/dx^2 < 0$ at $x = \frac{1}{3}S$, which means we have a maximum), which for UPS is acceptable, since then $x = \frac{1}{3} \cdot 130'' < 108''$. Thus, the cylindrical package of maximum volume has the volume

$$\pi \left(\frac{S - \frac{1}{3} S}{2\pi} \right)^2 \cdot \frac{S}{3} = \frac{S^3}{27\pi} = 0.0117893 \; S^3.$$

This is considerably larger than the spherical package of maximum volume, because the circular cross sections of a sphere do *not* all have the same (maximum) girth. We can see that this is so because, if the sphere has length (diameter) x then its radius is $\frac{1}{2}x$ and so its maximum girth is $2\pi \left(\frac{1}{2}x \right) = \pi x$. Thus, the UPS size is

$$x + \pi x = x(\pi + 1)$$

and the volume is

$$v = \frac{4}{3} \pi \left(\frac{1}{2}x \right)^3 = \pi \frac{x^3}{6}.$$

We clearly maximize v by simply maximizing x, which is achieved by dividing the maximum size S by $\pi + 1$. So, the volume of the largest mailable spherical package is

$$\pi \frac{\left(\frac{S}{\pi + 1} \right)^3}{6} = 0.0073705 \; S^3,$$

which is less than 63% the volume of the cylindrical package of maximum mailable volume.

Now, what if the cross sections of a package are all the same but are *not* necessarily circular? This results in a somewhat surprising conclusion. Let each identical cross section have area A and perimeter P. If we vary P (keeping the cross section *shape* fixed), then it is dimensionally clear that $A = kP^2$, where k is some positive constant ("depending" on the shape). Thus, if x is the package length, we have the package volume and size as

$$v = kP^2 x \quad \text{and} \quad S = x + P.$$

So,

$$x = S - P \quad \text{and} \quad v = kP^2(S - P) = kSP^2 - kP^3.$$

To find that P that maximizes v, we write

$$\frac{dv}{dP} = 0 = 2kSP - 3kP^2,$$

or $P = \frac{2}{3}S$. (Again, you should confirm that $d^2v/dP^2 < 0$ at $P = \frac{2}{3}S$, which means we have a maximum.) Thus, $x = S - P = \frac{1}{3}S$, *just as before* for a constant, *circular* cross section. That is, *independent* of the shape of the package's cross section, as long as it is the same everywhere, the package with maximum volume has length $\frac{1}{3}S$, one-third of the specified maximum size.

Finally, the paradox. Imagine a *cubical* package with edge length $5S/24$. Its size exceeds the maximum allowable because

$$\underbrace{\frac{5S}{24}}_{\text{length}} + 4 \cdot \underbrace{\frac{5S}{24}}_{\text{girth}} = \frac{25S}{24} > S,$$

and so this package is *not* mailable. But, it has a volume of

$$\left(\frac{5S}{24}\right)^3 = 0.0090422 \ S^3,$$

which is *significantly less* than the volume of the maximum volume cylindrical package. So, a UPS mail agent would accept the larger volume cylindrical package as small enough to mail, but would reject the smaller volume cubical package because it is too large! Ah, the complications of modern life.

5.4 Projectile Motion in a Gravitational Field

A classic use of the derivative is in the study of projectile motion through the Earth's gravitational field. In this section I'll first show a simple application of the derivative to a number of athletic events and then, in the next section, a related military example dating from 1686.

To start, imagine an athlete is a specialist in not only the shot put and the discus throw, but also in heaving the javelin *and* in golf! As different as these events are in their details, all can be expressed mathematically (at the most elementary level of sophistication) by a common model: the release of a projectile at height h above the ground, with initial speed v_0 at release, and at a release angle of θ. Eventually, the projectile returns to the Earth at some distance R from the point directly beneath the release point. The values of h and v_0 are assumed to be given for a given athlete; our problem here is to find the angle θ that maximizes R.

In the geometry of figure 5.4 (where the origin is the release point) we see that $y = -h$ when the projectile hits the ground at distance $x = R$. Using g to denote the acceleration of gravity, and realizing that only the vertical component of the projectile's speed is affected by gravity (I am ignoring any air-drag effect), we can write the horizontal and vertical components of the projectile's speed, at time t, as

$$\frac{dx}{dt} = v_0 \cos(\theta)$$

$$\frac{dy}{dt} = v_0 \sin(\theta) - gt, \qquad t \geq 0.$$

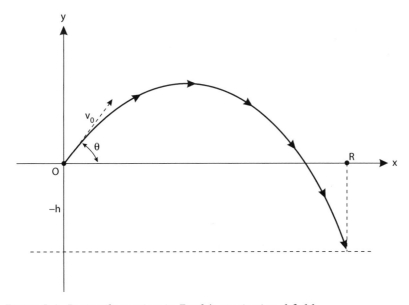

FIGURE 5.4. Projectile motion in Earth's gravitational field.

(Notice, carefully, that these two derivatives are present because of the *physics* of the problem, and not because of any extrema calculation.) Since $x(0) = y(0) = 0$, these two differential equations are easily integrated to give

$$x(t) = v_0 t \cos(\theta)$$

$$y(t) = v_0 t \sin(\theta) - \frac{1}{2} g t^2.$$

If we solve the first equation for t, i.e., if we write

$$t = \frac{x}{v_0 \cos(\theta)},$$

and then substitute into the second equation, we get

$$y = x \tan(\theta) - x^2 \frac{g}{2 v_0^2 \cos^2(\theta)}.$$

That is, y is a quadratic function of x, and so we have the well-known result that, for given values of v_0 and θ, the path of the projectile is a parabola.

Since $x = R$ when $y = -h$, then when the projectile hits the ground at time $t = \hat{t}$, we have

$$v_0 \hat{t} \cos(\theta) = R$$

$$v_0 \hat{t} \sin(\theta) - \frac{1}{2} g \hat{t}^2 = -h.$$

So, from the first equation,

$$\hat{t} = \frac{R}{v_0 \cos(\theta)},$$

and thus, from the second equation,

$$\frac{R \sin(\theta)}{\cos(\theta)} - \frac{1}{2} g \frac{R^2}{v_0^2 \cos^2(\theta)} = -h.$$

Or

$$R v_0^2 \cos(\theta) \sin(\theta) - \frac{1}{2} g R^2 + h v_0^2 \cos^2(\theta) = 0.$$

Using the trigonometric identity $\sin(2\theta) = 2\sin(\theta)\cos(\theta)$, this last expression becomes

$$\frac{1}{2} R v_0^2 \sin(2\theta) - \frac{1}{2} gR^2 + h v_0^2 \cos^2(\theta) = 0,$$

or, at last, a result so important I'll put it in a box:

$$\boxed{R v_0^2 \sin(2\theta) - gR^2 + 2h v_0^2 \cos^2(\theta) = 0.}$$

As our athlete's goal is to pick that θ (call it $\hat{\theta}$) that maximizes R, it now seems that we should introduce a derivative for *mathematical* reasons. Specifically, let's differentiate term-by-term with respect to θ using the result from section 5.1 for how to differentiate products:

$$R v_0^2 2 \cos(2\theta) + v_0^2 \sin(2\theta)\frac{dR}{d\theta} - 2gR\frac{dR}{d\theta} - 2h v_0^2 2 \cos(\theta)\sin(\theta) = 0.$$

At an extrema (a maximum of R), we will have $R'(\theta) = 0$, which gives

$$2R v_0^2 \cos(2\theta) - 4h v_0^2 \cos(\theta)\sin(\theta) = 0.$$

And finally, using the above double-angle identity once more, this reduces to

$$2R v_0^2 \cos(2\theta) - 2h v_0^2 \sin(2\theta) = 0,$$

or

$$R = h\tan(2\theta).$$

This isn't, however, quite what we are after, which is the particular θ that maximizes R for a given v_0 and h. But this result isn't useless, either; once we do have the value of that optimum $\theta = \hat{\theta}$, we can then find the actual distance achieved from $R_{\max} = h\tan(2\hat{\theta})$. But, first, what *is* $\hat{\theta}$? We can get our hands on $\hat{\theta}$ by substituting $R_{\max} = h\tan(2\hat{\theta})$ into our earlier boxed result that is true for *any* value of θ: $R v_0^2 \sin(2\theta) - gR^2 + 2h v_0^2 \cos^2(\theta) = 0$. Thus,

$$h v_0^2 \tan(2\hat{\theta})\sin(2\hat{\theta}) - gh^2\tan^2(2\hat{\theta}) + 2h v_0^2 \cos^2(\hat{\theta}) = 0,$$

or

$$v_0^2 \tan(2\hat{\theta}) \sin(2\hat{\theta}) + 2v_0^2 \cos^2(\hat{\theta}) = gh \tan^2(2\hat{\theta}),$$

or

$$v_0^2 \left[\frac{\sin^2(2\hat{\theta})}{\cos(2\hat{\theta})} + 2\cos^2(\hat{\theta}) \right] = gh \frac{\sin^2(2\hat{\theta})}{\cos^2(2\hat{\theta})}.$$

Since another trigonometric identity tells us that

$$\cos^2(\hat{\theta}) = \frac{1}{2}[1 + \cos(2\hat{\theta})],$$

this last result becomes

$$v_0^2 \left[\frac{\sin^2(2\hat{\theta})}{\cos(2\hat{\theta})} + 1 + \cos(2\hat{\theta}) \right] = gh \frac{1 - \cos^2(2\hat{\theta})}{\cos^2(2\hat{\theta})},$$

or

$$v_0^2 \left[\frac{\sin^2(2\hat{\theta}) + \cos(2\hat{\theta}) + \cos^2(2\hat{\theta})}{\cos(2\hat{\theta})} \right] = v_0^2 \frac{1 + \cos(2\hat{\theta})}{\cos(2\hat{\theta})}$$

$$= gh \frac{[1 + \cos(2\hat{\theta})][1 - \cos(2\hat{\theta})]}{\cos^2(2\hat{\theta})}.$$

So, after the obvious cancellation,

$$v_0^2 \cos(2\hat{\theta}) = gh[1 - \cos(2\hat{\theta})],$$

or, solving this easy equation for $\cos(2\hat{\theta})$,

$$\cos(2\hat{\theta}) = \frac{gh}{v_0^2 + gh} = \frac{g}{g + \dfrac{v_0^2}{h}} = \frac{g}{g + \alpha}, \qquad \alpha = \frac{v_0^2}{h}.$$

The parameter α is characteristic of each particular athlete, depending on both height h and strength (the speed v_0 of the projectile at the instant of release). So now, at last, we have the optimum value of θ:

$$\hat{\theta} = \frac{1}{2}\cos^{-1}\left\{\frac{g}{g+\alpha}\right\}$$

to give

$$R_{max} = h\tan\left(2\hat{\theta}\right).$$

An exceptional case occurs for golf. There, h is not a height associated with the player, as in the track-and-field events of the shot put, the javelin throw, and the discus toss. Rather, h is the height of the ball tee, which I'll take as essentially zero. Thus, independent of v_0 we have $\alpha = \infty$ and so $\hat{\theta} = \frac{1}{2}\cos^{-1}(0) = 45°$, i.e., all golfers, independent of their individual strengths, have the same optimal angle when swinging for distance. The actual distance achieved *does* of course, depend greatly on v_0.

Interestingly, for golf, our result for R_{max},

$$R_{max} = h\tan(2\hat{\theta}),$$

is indeterminate (useless) because it reduces to

$$R_{max} = 0 \cdot \infty = ???$$

To determine R_{max} for the golf case of $h = 0$, let's return to our earlier boxed result just *before* we differentiated with respect to θ, which is true for any θ:

$$Rv_0^2\sin(2\theta) - gR^2 + 2hv_0^2\cos^2(\theta) = 0.$$

For $\theta = \hat{\theta} = 45°$, we have $R = R_{max}$, and as $h = 0$, then

$$R_{max}v_0^2 - g\,R_{max}^2 = 0,$$

or

$$R_{max} = \frac{v_0^2}{g}.$$

A strong golfer can drive a ball off its tee at about $v_0 = 160$ feet/second, and so this analysis predicts the maximum driving distance to be

$$R_{max} = \frac{(160 \text{ ft/sec})^2}{32.2 \text{ ft/sec}^2} = 795 \text{ ft.}$$

This is, indeed, a long drive, but one that occasionally is actually observed.

Now, one last point. A physicist or engineer would argue that it is *physically* obvious that our result for $\hat{\theta}$ gives a maximum in R, not a minimum. For $\theta > \hat{\theta}$, the projectile spends most of its time traveling vertically, not horizontally. For $\theta < \hat{\theta}$, gravity pulls the projectile back to Earth "too soon." A mathematician, however, would want to apply the second derivative test, and this is a good exercise for *you* to run through. Simply start with the result of the first differentiation (*before* we set $R'(\theta) = 0$) and differentiate it again. *Then* set $R'(\theta) = 0$, as well as use our two results for that optimal case $R = h \tan(2\theta)$ and $\cos(2\theta) = gh/(v_0^2 + gh)$). That will result (if you are careful with the algebra) in $R''(\theta) < 0$, which means the extrema in R is, indeed, a maximum.

5.5 The Perfect Basketball Shot

An interesting (and historically important, as you'll soon see) twist to the analysis of the previous section can also be found in a non-track-and-field athletic event: basketball. Assume a player is standing directly in front of a basketball hoop, preparing to make a free-throw shot. If we construct a coordinate system with its origin at the release point (i.e., where the ball leaves the player's hands), then we have the geometry shown in figure 5.5. The ball is released at time $t = 0$ from the origin, at a launch angle θ, with initial speed v_0, with the goal of having the ball drop through the hoop, i.e., of having the ball pass through the hoop's location at $(x = \ell, y = h)$ on the *falling* portion of its parabolic trajectory. In this section, I'll show you how calculus lets us determine the minimum value of v_0 required to do

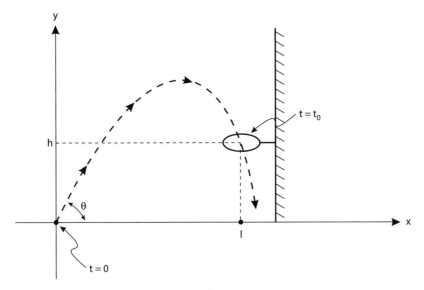

FIGURE 5.5. Geometry of basketball shooting.

this, and then I'll explain how (and why) this problem was stated and solved more than *three centuries* ago, long before the invention of basketball. (The explanation does *not* involve time travel!) Much of what follows was inspired by an essay written by C. W. Groetsch, "Halley's Gunnery Rule" (*The College Mathematics Journal*, January 1997, pp. 47–50).

If we say that the ball passes through the hoop at time $t = t_0$, then from the previous section we know that $x(t_0) = \ell$ and $y(t_0) = h$, where

$$x(t) = v_0 t \cos(\theta), \quad y(t) = v_0 t \sin(\theta) - \frac{1}{2} g t^2.$$

That is, we require

$$\ell = v_0 t_0 \cos(\theta), \quad h = v_0 t_0 \sin(\theta) - \frac{1}{2} g t_0^2,$$

and so

$$t_0 = \frac{\ell}{v_0 \cos(\theta)},$$

and therefore,

$$h = \ell \frac{\sin(\theta)}{\cos(\theta)} - \frac{1}{2} g \frac{\ell^2}{v_0^2 \cos^2(\theta)}.$$

That is,

$$h = \ell \tan(\theta) - \frac{1}{2} \frac{g\ell^2}{v_0^2} \sec^2(\theta).$$

Solving for v_0^2, we arrive at the somewhat complicated looking result,

$$v_0^2 = \frac{\frac{1}{2} g\ell^2 \sec^2(\theta)}{\ell \tan(\theta) - h},$$

which tells us with what initial speed the player must send the ball on its way, given the hoop location (the values of h and ℓ) and the launch angle (θ).

We can now derive an interesting mathematical constraint on v_0^2 that shows there is a minimum initial speed if the ball is to pass through the point (ℓ, h). This makes sense *physically*, of course. After all, if the loop is (for example) 25 feet (horizontally) away from the player, and the hoop is 10 feet above the court, then even a nonmathematician, couch-potato, Larry Bird wanna-be knows intuitively that the puny launch speed of $v_0 = 1$ foot/second isn't going to score any points! With some simple algebra we can find an expression for the minimum launch speed, in terms of ℓ and h. (This is equivalent to asking for the minimum *energy* shot.)

Since $\sec^2(\theta) = 1 + \tan^2(\theta)$, then

$$v_0^2 = \frac{\frac{1}{2} g\ell^2 [1 + \tan^2(\theta)]}{\ell \tan(\theta) - h},$$

or

$$\frac{1}{2} g\ell^2 + \frac{1}{2} g\ell^2 \tan^2(\theta) = v_0^2 \ell \tan(\theta) - v_0^2 h,$$

or

$$\frac{1}{2}g\ell^2 \tan^2(\theta) - v_0^2\ell \tan(\theta) + \frac{1}{2}g\ell^2 + v_0^2 h = 0,$$

or, finally,

$$\tan^2(\theta) - \frac{2v_0^2}{g\ell} \tan(\theta) + 1 + \frac{2v_0^2 h}{g\ell^2} = 0,$$

a quadratic in $\tan(\theta)$. So, using the quadratic formula to solve for $\tan(\theta)$, we have a result so important I'll put it in a box:

$$\tan(\theta) = \frac{v_0^2}{g\ell} \pm \frac{1}{2}\sqrt{\frac{4v_0^4}{g^2\ell^2} - 4 - \frac{8v_0^2 h}{g\ell^2}}.$$

For this to make physical sense we demand that $\tan(\theta)$ be real, i.e., that the square root be of a nonnegative quantity. So,

$$\frac{4v_0^4}{g^2\ell^2} - 4 - \frac{8v_0^2 h}{g\ell^2} \geq 0,$$

which is easily manipulated into

$$v_0^4 - 2ghv_0^2 - g^2\ell^2 \geq 0.$$

The left-hand side of this inequality is a quadratic in v_0^2, to which we can again apply the quadratic formula to conclude that

$$v_0^2 \geq gh \pm g\sqrt{h^2 + \ell^2}.$$

But we can immediately reject the negative root because v_0^2 must, of course, be positive. Thus, we write

$$v_0^2 \geq g\left(h + \sqrt{h^2 + \ell^2}\right).$$

From our earlier numerical example of $\ell = 25$ feet and the hoop 10 feet above the court, then $h = 4$ feet for a player who releases the ball at a height 6 feet above the court, and we have

$$v_0^2 \geq 32.2\left(4 + \sqrt{16 + 625}\right)\frac{\text{ft}^2}{\sec^2} = 944\frac{\text{ft}^2}{\sec^2}.$$

For the ball to pass through the hoop at (25,4) we must have $v_0 \geq$ 30.7 feet/second, i.e., the *minimum* speed is $v_0 = 30.7$ feet/second. This result does *not*, however, completely define what the player has to do to score with minimum expended energy; he must also determine the launch angle.

To find the angle of the minimum energy shot, return to the boxed $\tan(\theta)$ expression and use the fact that *at* minimum launch speed the quartic inequality for v_0 becomes an equality $(v_0^4 - 2ghv_0^2 - g^2\ell^2 = 0)$, and so

$$\tan(\theta) = \frac{v_0^2}{g\ell},$$

or

$$\theta = \tan^{-1}\left(\frac{v_0^2}{g\ell}\right) = \tan^{-1}\left[\frac{944}{(32.2)(25)}\right] = 49.54°.$$

But we *still* are not quite done. We have, so far, not specifically imposed the requirement that the ball *drop* through the hoop (the other way for the ball to pass through the hoop is on the *upward* portion of its trajectory, which is clearly *not* a legal basketball play!) We need to explore this issue next.

The mathematical requirement we need, at time $t = t_0$, is that

$$\left.\frac{dy}{dt}\right|_{t=t_0} < 0,$$

which is simply the requirement that the ball's vertical speed be *negative* as the ball passes through the hoop, i.e., *downward*-directed toward the ground. That insures that the ball is *falling* through the hoop. Thus, as

$$\frac{dy}{dt} = v_0\sin(\theta) - gt$$

in general, then at time $t = t_0$, we can write

$$v_0\sin(\theta) - gt_0 < 0.$$

This says

$$v_0 < \frac{g t_0}{\sin(\theta)} = \frac{g \dfrac{\ell}{v_0 \cos(\theta)}}{\sin(\theta)} = \frac{g\ell}{v_0 \sin(\theta) \cos(\theta)}.$$

That is,

$$v_0^2 < \frac{g\ell}{\sin(\theta) \cos(\theta)}.$$

Combining this with our previous result for v_0^2, we have

$$\frac{\frac{1}{2} g\ell^2 \sec^2(\theta)}{\ell \tan(\theta) - h} < \frac{g\ell}{\sin(\theta) \cos(\theta)}.$$

Dividing through by $g\ell$ and then cross-multiplying, this becomes

$$\frac{1}{2} \ell \sin(\theta) \cos(\theta) \sec^2(\theta) < \ell \tan(\theta) - h,$$

or, since $\sin(\theta) \cos(\theta) \sec^2(\theta) = \tan(\theta)$, we have

$$\frac{1}{2} \ell \tan(\theta) < \ell \tan(\theta) - h.$$

Thus,

$$h < \frac{1}{2} \ell \tan(\theta)$$

and so, for the ball to *drop* through the hoop, we have the following inequality that must be satisfied by the launch angle:

$$\theta > \tan^{-1}\left(\frac{2h}{\ell}\right).$$

The question now is: what angle θ goes with the minimum velocity, i.e., does the θ associated with the minimum energy shot satisfy the above inequality? If it does, then the ball does indeed *drop* through the hoop. Otherwise, the ball must *rise* through the hoop and that would, in the context of our original problem, be an illegal shot. So, as I just did for the specific numerical example, let's return to the boxed $\tan(\theta)$ equation but now insert the general *expression* for the minimum v_0^2. As in the numerical example, the square root in the boxed $\tan(\theta)$ equation is zero and so the required launch angle is

$$\tan(\theta) = \frac{v_0^2}{g\ell} = \frac{g\left(h + \sqrt{h^2 + \ell^2}\right)}{g\ell} = \frac{h + \sqrt{h^2 + \ell^2}}{\ell}$$

$$= \frac{h}{\ell} + \sqrt{\left(\frac{h}{\ell}\right)^2 + 1} > 2\left(\frac{h}{\ell}\right).$$

Thus, the answer to our question is *yes*, if the minimum launch speed is used, then the condition on θ, for the ball to *drop* through the hoop, *is* satisfied.

The minimum speed (minimum energy) launch angle has a very interesting geometric interpretation. In figure 5.6, I have constructed a right triangle with a base angle of θ, by giving it a base length of

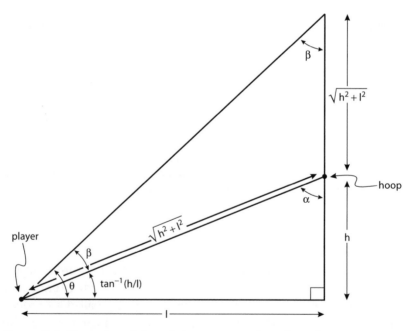

FIGURE 5.6. Geometry of the minimum energy launch angle.

ℓ and then two consecutive components to the vertical side; one of length h and the other of length $\sqrt{h^2 + \ell^2}$. This right triangle has then been divided into two other triangles, which I'll call the upper and lower triangles. Since the hypotenuse of the lower triangle (which is also a side of the upper triangle) has length $\sqrt{h^2 + \ell^2}$, then the upper triangle is isosceles, which is why I have given the same angle β to the two angles shown in figure 5.6. The last angle labeled is α, the top angle of the lower triangle.

From elementary geometry we can now write the following sequence of statements:

(a) $\theta + \beta = 90°$, or $\theta = 90° - \beta$;
(b) $\alpha + \tan^{-1}(h/\ell) = 90°$,   or $\alpha = 90° - \tan^{-1}(h/\ell)$;
(c) $2\beta + (180° - \alpha) = 180°$,   or $\beta = \frac{1}{2}\alpha = 45° - \frac{1}{2}\tan^{-1}(h/\ell)$.

Substituting the expression for β into the expression for θ in (a), we have

$$\theta = 90° - \left[45° - \frac{1}{2}\tan^{-1}\left(\frac{h}{\ell}\right)\right] = 45° + \frac{1}{2}\tan^{-1}\left(\frac{h}{\ell}\right)$$

$$= \frac{90° + \tan^{-1}\left(\frac{h}{\ell}\right)}{2}.$$

That is, the minimum-launch-energy shot has a launch angle given by the *average* of the line-of-sight angle from the player to the hoop, and the vertical. For example, returning to our numerical example of $\ell = 25$ feet and $h = 4$ feet, the line-of-sight angle to the hoop is $\tan^{-1}(4/25) = 9.09°$, and so the launch angle for the minimum speed (minimum energy) shot is

$$\theta = \frac{90° + 9.09°}{2} = 49.54°,$$

just as we calculated earlier in the numerical example.

5.6 Halley's Gunnery Problem

The basketball problem was originally solved in 1686, and it appeared in print in a paper published in 1688 by the Royal Society in its *Philosophical Transactions*. The author (who as Editor was easily able to arrange to have the publication back-dated to 1686), was Edmond Halley (1656–1742), and he was obviously motivated by something other than basketball, of course, as that activity didn't arrive on the scene until considerably later. Today we remember Halley mostly for two reasons; the comet named after him because he was the first to recognize it as a periodically returning body traveling on a greatly elongated elliptical orbit around the sun, and for being the force (both spiritually and financially) behind getting Newton's masterpiece *Principia Mathematica* published in 1687. (Halley was also the "infidel"—because he had convinced a mutual acquaintance that the Christian faith is a fairy tale—mentioned in the subtitle of Bishop Berkeley's attack on the logic of calculus, *The Analyst*.) But Halley was also an accomplished scientist and mathematician in his own right, and his solution to the "basketball problem" shows a first-class intellect at work.

The last phrase of the rather long title to Halley's paper gives us a clue to his motivation: "A discourse concerning gravity, and its properties wherein the descent of heavy bodies, and the motion of projectiles is briefly but fully handled: together with the solution of a problem of great use in gunnery." What Halley did in this paper was to address the problem of determining the best way for a cannon to lob a projectile onto a target located above the gun, e.g., onto a town high up on a mountain side, with the gun located in the plains far below. As Halley wrote, in a second paper published in 1695 (which contains a derivation of the minimum speed launch angle as the average of the line of sight and the vertical angles), it isn't a good idea to simply blast away with all of the power the gun could possibly provide. That's because such energetic projectiles arrive on target at such high speed that they "bury themselves too deep in the ground, to do all the damage that they might . . . which is a thing acknowledged by the besieged in all towns, who unpave their streets, to let the bombs bury themselves and thereby stifle the force of their splinters."

Halley therefore reasoned that the proper way to launch a bomb at the higher elevation target was to arrange for the bomb to drop onto the target with minimum kinetic energy. Now, even though a cannon-fired projectile is moving through the air at speeds much faster than a shot, a discus, a basketball, or even a golf ball, Halley did as I have done in the basketball analysis, and ignored all air-drag effects. That is, we will continue to assume energy is conserved, and so the sum of the kinetic and potential energies of the projectile will, at every instant of time, be a constant.

At launch, the projectile has only kinetic energy of motion, and zero potential energy. At impact, it has the potential energy due to the height of the target, plus the kinetic energy of its motion at impact (which should be as small as possible). There is, of course, nothing we can do about the potential energy at the target height, and so to minimize the impact kinetic energy, one must minimize the launch (kinetic) energy, i.e., minimize the launch speed. And so Halley arrived at the basketball problem, long before the invention of basketball. An immediate implication of this conclusion is that the powder charge needed to send the projectile on its way is also minimized. This was, no doubt, attractive to those responsible for how the king's coin was spent. The immediate question *this* raises, of course, is just what *is* the powder charge required to deliver a projectile, with minimum energy, to an elevated target? Halley answered this question, too.

As derived in the golf ball example of section 5.4, a ball driven off of its tee at an initial speed of v_0, *at an angle of 45°*, achieves its maximum horizontal range of v_0^2/g over a horizontal surface. What is true for the golf ball is true for the cannon projectile (ignoring air drag), *if* the cannon is fired over a horizontal surface with its barrel elevated to 45°. Since $v_0^2 = g(h + \sqrt{h^2 + \ell^2})$ for the minimum energy shot, then the value of v_0^2/g is $h + \sqrt{h^2 + \ell^2}$, and this gives us Halley's so-called *calibration rule*: the powder charge required to deliver a projectile onto a target at (ℓ, h) with minimum kinetic energy is the same charge required to shoot the same projectile out of the cannon (with 45° of barrel elevation) to a distance of $h + \sqrt{h^2 + \ell^2}$. A series of test firings for any given cannon and projectile could give a table of powder charge versus projectile range. It is clear, of course, that it is possible to have two targets, with very different values of ℓ and h, requiring the same powder charge. For example, a target at (2000, 1000) has the same required powder charge as a target at (2690, 500). All that remained for the gunner to do, then, was to use Halley's angle rule to get the proper elevation of the cannon barrel. For our two targets, for example, the elevation angles are, respectively,

$$\frac{90° + \tan^{-1}\left(\dfrac{1{,}000}{2{,}000}\right)}{2} = 58.3°$$

and

$$\frac{90° + \tan^{-1}\left(\dfrac{500}{2{,}690}\right)}{2} = 50.3°.$$

The proper barrel elevation angle has a special minimization property that Halley also discovered, in response to a very practical concern. Suppose the gunner makes a slight error in setting the elevation angle. How would that affect the accuracy of the bombardment? That is, if he makes a slight change (error) of $\Delta\theta$ from the correct θ, then how much of a change is made in the impact point? Note carefully that we are making an error only in θ; the powder charge, and hence v_0, is taken as correct.

We start by recalling a result from the previous section,

$$\tan^2(\theta) - \frac{2v_0^2}{g\ell}\tan(\theta) + 1 + \frac{2v_0^2 h}{g\ell^2} = 0.$$

Remember what these symbols mean: h is the height of the target, and thus is a constant, but ℓ (the range of the projectile when it is at height h) depends on θ. If ℓ is the range to the target, *then*, by definition, θ is set correctly because then the projectile and target coincide! To simplify the algebra, let's make the following definitions:

$$u = \tan(\theta), \quad \text{a variable;}$$

$$a = \frac{h}{\ell}, \quad \text{a variable;}$$

$$p = \frac{2v_0^2}{g}, \quad \text{a constant.}$$

Then,

$$u^2 - \frac{p}{\ell}u + 1 + p\frac{a}{\ell} = 0,$$

which is easily solved for ℓ:

$$\ell = \frac{p(u-a)}{u^2+1}.$$

To find how ℓ varies with small changes in θ, we can use the chain rule. From the very definition of the derivative, if $\Delta\theta$ is a "small" change in θ, then the change in ℓ is $\Delta\ell$, where

$$\Delta\ell \approx \Delta\theta \, \frac{d\ell}{d\theta}.$$

By the chain rule,

$$\frac{d\ell}{d\theta} = \frac{d\ell}{du} \cdot \frac{du}{d\theta} = \frac{d\ell}{du} \cdot \frac{d}{d\theta}[\tan(\theta)] = \frac{1}{\cos^2(\theta)} \cdot \frac{d\ell}{du}.$$

Now,

$$\frac{d\ell}{du} = \frac{d}{du}\left[\frac{p(u-a)}{u^2+1}\right] = p\frac{d}{du}\left[\frac{u-a}{u^2+1}\right]$$

$$= p\frac{(u^2+1)\left(1-\dfrac{da}{du}\right) - (u-a)2u}{(u^2+1)^2}$$

$$= p\frac{u^2+1-2u^2+2au - (u^2+1)\dfrac{da}{du}}{(u^2+1)^2}$$

$$= p\frac{1+2au-u^2 - (u^2+1)\dfrac{da}{du}}{(u^2+1)^2}.$$

Remembering that h is a constant, we have

$$\frac{da}{du} = \frac{d}{du}\left(\frac{h}{\ell}\right) = \frac{\ell\dfrac{dh}{du} - h\dfrac{d\ell}{du}}{\ell^2} = -\frac{h}{\ell^2} \cdot \frac{d\ell}{du} = -\frac{a}{\ell} \cdot \frac{d\ell}{du},$$

and so

$$\frac{d\ell}{du} = p\frac{1+2au-u^2 + (u^2+1)\dfrac{a}{\ell} \cdot \dfrac{d\ell}{du}}{(u^2+1)^2}.$$

This can be solved for $d\ell/du$ to give

$$\frac{d\ell}{du} = \left\{\frac{p\ell}{\left[\ell(u^2+1) - ap\right](u^2+1)}\right\}\left\{(1 + 2au - u^2)\right\}.$$

The two factors on the right-hand side are such that the second one is zero and the first one is finite. To see this, consider the following.

When the gunner's aim is perfect, i.e., when he has set the angle θ so that

$$\tan(\theta) = u = \frac{h}{\ell} + \sqrt{\left(\frac{h}{\ell}\right)^2 + 1} = a + \sqrt{a^2 + 1},$$

(as shown in the previous section), then we have

$$1 + 2au - u^2 = 1 + 2a\left[a + \sqrt{a^2+1}\right] - \left[a + \sqrt{a^2+1}\right]^2$$

$$= 1 + 2a^2 + 2a\sqrt{a^2+1} - \left[a^2 + 2a\sqrt{a^2+1} + a^2 + 1\right]$$

$$= 0.$$

Thus, the second factor of $d\ell/du$ vanishes.

For the first factor of $d\ell/du$, notice that since $u^2 - (pu/\ell) + 1 + p(a/\ell) = 0$, then $\ell u^2 + \ell = pu - ap$. Thus,

$$1 + \ell(u^2 + 1) - ap = \ell u^2 + \ell - ap = pu - ap - ap = pu - 2ap,$$

and so the first factor of $d\ell/du$ is proportional to

$$\frac{p\ell}{pu - 2ap} = \frac{\ell}{u - 2a} = \frac{\ell}{a + \sqrt{a^2+1} - 2a} = \frac{\ell}{\sqrt{a^2+1} - a} > 0,$$

i.e., the first factor is positive and, more importantly, *finite*. Thus, when the gunner's aim is perfect we see that

$$\frac{d\ell}{du} = 0,$$

which says

$$\frac{d\ell}{d\theta} = \frac{1}{\cos^2(\theta)} \cdot \frac{d\ell}{du} = 0.$$

Thus, when the angle is set correctly, we have $d\ell/d\theta = 0$ and this says that, even when the aim is set *not* so perfect but is still "in the neighborhood" around perfect aim, we have

$$\Delta\ell \approx \Delta\theta \frac{d\ell}{d\theta} \approx 0.$$

Halley summarized his minimum results, including this last one, as follows: "This Rule may be of good use to all Bombardiers and Gunners, not only that they may use no more Powder than is necessary, to cast their Bombs into the place assigned, but that they may shoot with much more certainty, for that a small Error committed in the Elevation of the Piece, will produce no sensible difference in the fall of the Shot." Thus wrote Edmond Halley over three centuries ago, one of the world's first modern theoreticians in the arcane art of military weapons analysis.

5.7   De L'Hospital and His Pulley Problem, and a New Minimum Principle

It is a curious fact that while Newton's *Principia* is the origin of modern physics, a subject universally presented to modern students using Newton's co-invention of the calculus, *Principia* itself does *not* use calculus. Rather, Newton presented the new physics with the mathematical aid of the "old math," Euclidean geometry, which makes for a presentation vastly more difficult than does the modern approach. Why did Newton do that, even though he obviously possessed the math we use today (he invented it!)? Almost surely the answer is that Newton wanted to avoid distracting his readers from the *physics*, which the then still mysterious calculus would have done. In *Principia*, Newton's goal was to champion his physics, not his math.

The recognition for publishing the world's first calculus book, then, goes not to Newton but to another. The French mathematician Guillaume-Francois-Antoinė de L'Hospital (1661–1704), who was an army cavalry officer until bad eyesight caused him to resign, has that honor. He was a quick study who could readily absorb the discoveries of others and then present them in a coherent manner for a wide audience. The contemplation of mathematics, then, was the perfect activity for a bright but nearsighted gentleman.

In de L'Hospital's time, when elementary calculus techniques were first being discovered, there weren't a lot of pedagogical resources around. So, what he did was hire the brightest of Leibniz's own students, Johann Bernoulli (1667–1748), then still a young man, to teach him the new math. (We'll hear again from Bernoulli, in the next chapter, in connection with one of the most famous minimization problems in mathematics.) De L'Hospital paid Bernoulli well and came to believe that, since he *had* paid for the new results, then those accomplishments were *his*. Such a "purchase" of intellectual property rights would, today, be considered acceptable only in matters like a celebrity hiring a ghostwriter to pen a so-called autobiography, an activity that is itself on the borderline of dubious authorship.

By 1696, de L'Hospital felt he had sufficient material on hand from Bernoulli to publish a book, *Analyse des Infiniment Petits* (analysis of the infinitely small). While containing nothing of his own discoveries, the book was well written and quickly became famous. It did contain, however, many of *Bernoulli's* discoveries, as well as those of Newton, Leibniz, and Bernoulli's older brother Jacob. Some writers have commented that de L'Hospital gave no mention at all to the Bernoulli brothers, and others have said that he did. In fact, he did—but not very much! Two brief sentences appear in the preface—"I am obliged to the gentlemen Bernoulli for their many bright ideas; particularly to the younger Mr. Bernoulli who is now a professor. I have made free use of their discoveries. . . ." In fact, de L'Hospital's words are a vast understatement. After de L'Hospital's death, Bernoulli began to claim credit for nearly all of the book, a claim initially rejected by mathematicians and historians alike. However, in 1922, a copy of a set of notes taken during a series of lectures Bernoulli gave on the differential calculus in Geneva, in 1691, was discovered. The organization and content of those notes, written *five years* before de L'Hospital's book, are virtually identical with the book.

_____

The classic example of de L'Hospital's appropriation of Bernoulli's work is the famous rule for calculating indeterminate limits. It is often the case that an analyst needs to compute the value of a ratio of two functions of the same independent variable (call it x) as x approaches zero. That is, she needs to compute the limit

$$R = \lim_{x \to 0} R(x) = \lim_{x \to 0} \frac{g(x)}{h(x)}.$$

Often, this is an easy calculation. For example, it is clear that

$$R = \lim_{x \to 0} R(x) = \lim_{x \to 0} \frac{3x + 8}{2x + 4} = \frac{8}{4} = 2.$$

But what do we do with something like

$$R = \lim_{x \to 0} R(x) = \lim_{x \to 0} \frac{\sin(x)}{x},$$

which reduces to the indeterminate $0/0$ if we simply stick $x = 0$ into the numerator and the denominator of the ratio? While it was *Bernoulli* who showed

$$R = \lim_{x \to 0} R(x) = \lim_{x \to 0} \frac{g'(x)}{h'(x)},$$

and so

$$R = \lim_{x \to 0} R(x) = \lim_{x \to 0} \frac{\sin(x)}{x} = \lim_{x \to 0} \frac{\cos(x)}{1} = \lim_{x \to 0} \cos(x) = 1,$$

this formula is instead known today as *L'Hospital's rule*, not as Bernoulli's rule. It is easy to derive.

Since $g(x) = R(x)h(x)$, then differentiation of both sides gives

$$g'(x) = R(x)h'(x) + R'(x)h(x).$$

Since $\lim_{x \to 0} h(x) = 0$, and if we assume $R(x)$ really does have a limit as $x \to 0$, i.e., $\lim_{x \to 0} R(x) = R$, then

$$\lim_{x \to 0} g'(x) = \lim_{x \to 0} R(x) \, h'(x) + \lim_{x \to 0} R'(x) \, h(x)$$

$$= R \lim_{x \to 0} h'(x) + \lim_{x \to 0} R'(x) \lim_{x \to 0} h(x).$$

The last term is zero because $\lim_{x \to 0} h(x) = 0$ *and* because the very fact that $\lim_{x \to 0} R(x) = R$ implies that $\lim_{x \to 0} R'(x) = 0$, too (i.e., the $y = R(x)$ curve must approach the *horizontal*, zero-*slope* line $y = R$ as $x \to 0$). So,

$$\lim_{x \to 0} g'(x) = R \lim_{x \to 0} h'(x),$$

and we have L'Hospital's rule.

Oddly enough, de L'Hospital was actually quite a *good* mathematician in his own right, and so why he did what he did remains, I think, a bit of a puzzle. Bernoulli, for his part, had remained silent until 1704 because his agreement with de L'Hospital had been that, in exchange for the rather large payments made for giving his math lessons, Bernoulli *would* remain quiet. Bernoulli's own complicity in this peculiar "contract" is also perplexing. Well, no matter the conflicted ethical issues involved with de L'Hospital's book, it does contain a number of interesting mathematical problems. One of them, in particular, demonstrates not only the differential calculus, but also the power of a new minimum principle similar in spirit to Fermat's least-time principle (see Alexander J. Hahn, "Two Historical Applications of Calculus," *The College Mathematics Journal*, March 1998, pp. 93–103).

The problem, easy to visualize, is illustrated in figure 5.7. At point A on a ceiling we attach one end of an idealized, weightless cable of length r. The other end of this completely flexible cable is attached to the center axle of a weightless pulley. At point B on the ceiling, distance d from A, we attach one end of another idealized cable of length ℓ, and pass it over the pulley. The other end of this second cable is attached to a block of material with weight W. We then let this system of cables, pulley, and weight freely adjust itself to its final, stationary (unmoving) configuration under the influence of gravity, with the pulley's ultimate location labeled as point C.

If $r < d$, then it is physically clear that the final equilibrium position of the system will be as shown in the illustration, with C below and between A and B, and with the weight directly below C, at point D. This is the case of *mathematical* interest, as well,

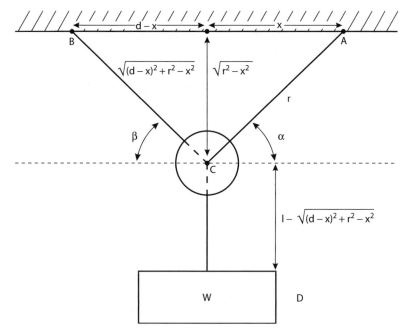

FIGURE 5.7. Geometry of L'Hospital's pulley problem.

because if $r > d$, it is equally obvious that then the weight would simply hang straight below B. That is, the weightless pulley would slide along the cable attached to the weight until the weightless pulley cable is pulled straight (or, if $r > \sqrt{\ell^2 + d^2}$, until the pulley rests on top of the weight). Therefore, if $r > d$, the weight hangs directly beneath B, distance ℓ below the ceiling, and that completely describes the equilibrium configuration of the system.

Far more interesting, physically *and* mathematically, is the case $r < d$. What, then, is the equilibrium configuration of the system? That is, where does the pulley end up? To start the analysis of this question, let point C be distance x to the left of A (and so distance $d - x$ to the right of B). Obviously, $0 \le x \le r$. Using the Pythagorean theorem twice, it is easy to see, as illustrated in figure 5.7, that the distance the weight hangs below the ceiling is the function x (let's call it $f(x)$) given by

$$f(x) = \sqrt{r^2 - x^2} + \ell - \sqrt{(d - x)^2 + r^2 - x^2}.$$

It may not be obvious, however, just what we should *do* with $f(x)$ to help us find where C is. In fact, to continue with the calculus solution I now need to introduce that new minimum principle I mentioned earlier. Before I do that, however, let me solve the problem in an entirely different way, not using calculus, and then when we return to $f(x)$ and apply calculus to it, we will be able to check the answer (be assured, the answers *will* agree!) The calculus approach will prove to be the easier to perform.

De L'Hospital's pulley problem is actually a type of problem commonly encountered by students in a first-year course in physics and engineering, when studying *statics* (the physics of *unmoving* systems). The key physical observation is that the cables, pulley, and weight are not moving in their final, stable configuration. In particular, the pulley is not moving. Newton's physics then tells us that this means there is no *net* force acting on the pulley; otherwise the pulley would be accelerated, i.e., it *would* move. So, the stable, or equilibrium, configuration can be found by setting the sum of all of the individual horizontal forces acting on the pulley to zero, and similarly for the sum of all the individual vertical forces acting on the pulley. Those forces come from the tensions in the two cables. The cable attached to the weight has tension W. This is clearly so for the vertical portion of that cable and, since the tension must be the same all along the cable (if not, there would be some place on the cable with a nonzero net force there and that part of the cable would move, contrary to the reality that the cable is *not* moving) the tension is everywhere W, even in the nonvertical portion of the cable. The horizontal component of this tension is directed to the left, with value $W \cos(\beta)$. If we call the tension in the other cable (the one attached to the pulley) T, then that tension has horizontal component $T \cos(\alpha)$ directed to the right. Thus,

$$W \cos(\beta) - T \cos(\alpha) = 0,$$

or

$$T = W \frac{\cos(\beta)}{\cos(\alpha)}.$$

Now, the vertical sum of forces on the pulley is given by

$$W \sin(\beta) + T \sin(\alpha) - W = 0.$$

Substituting in for T,

$$W\sin(\beta) + W\frac{\cos(\beta)\sin(\alpha)}{\cos(\alpha)} - W = 0,$$

or

$$\sin(\beta) + \cos(\beta)\tan(\alpha) - 1 = 0.$$

From the geometry of figure 5.7, we can write

$$\sin(\beta) = \frac{\sqrt{r^2 - x^2}}{\sqrt{(d-x)^2 + r^2 - x^2}},$$

$$\cos(\beta) = \frac{d-x}{\sqrt{(d-x)^2 + r^2 - x^2}},$$

$$\tan(\alpha) = \frac{\sqrt{r^2 - x^2}}{x}.$$

Thus,

$$\frac{\sqrt{r^2 - x^2}}{\sqrt{(d-x)^2 + r^2 - x^2}} + \frac{d-x}{\sqrt{(d-x)^2 + r^2 - x^2}} \cdot \frac{\sqrt{r^2 - x^2}}{x} - 1 = 0,$$

which can be algebraically manipulated into

$$2x^2 d - r^2 x - r^2 d = 0.$$

This quadratic in x is now easy to solve (and I'll do that in just a bit), and our question (where is point C, the location of the pulley?) is answered.

Now, let's return to that $f(x)$ function derived earlier (which tells us how far below the ceiling the weight hangs). The new minimum principle I mentioned before is now applied—the system is in stable equilibrium when its *potential energy is minimum*. (We'll use this same argument again, in the next chapter, to solve a much more famous problem than this one.) That is, stable equilibrium occurs when the weight hangs as far below the ceiling as possible, which occurs when $f(x)$ is *maximum*. So, all we need to do is set the derivative of $f(x)$ equal to zero and solve for x, i.e.,

$$\frac{df}{dx} = \frac{-2x}{2\sqrt{r^2 - x^2}} - \frac{-2(d-x) - 2x}{2\sqrt{(d-x)^2 + r^2 - x^2}} = 0.$$

Once again, this expression is easy to algebraically manipulate to give

$$2x^2 d - r^2 x - r^2 d = 0,$$

precisely the quadratic result we got from the statics analysis.

To finish the problem, all we need do now is to actually solve the quadratic. We, of course, formally get two answers:

$$x = \frac{r^2 \pm \sqrt{r^4 + 8d^2 r^2}}{4d} = \frac{r}{4d}\left[r \pm \sqrt{r^2 + 8d^2}\right].$$

It is physically obvious that x is not negative, and so we reject the negative root. So, the location of the pulley is given by

$$x = \frac{r}{4d}\left[r + \sqrt{r^2 + 8d^2}\right].$$

Notice, too, that the constraint $x < r$ is also satisfied, because we can write x as

$$x = \frac{1}{4} r \left[\frac{r}{d} + \sqrt{\left(\frac{r}{d}\right)^2 + 8}\right]$$

and, since $r/d < 1$, it follows that

$$x < \frac{1}{4} r \left[1 + \sqrt{1 + 8}\right] = r.$$

As a final comment on de L'Hospital's pulley problem, notice that the solution value of x has no dependence on either the weight W, or on the length ℓ of cable attached to the weight. (For many people, including me, this is *not* intuitively obvious!) Only the length of the cable connected to the pulley (r), and distance between the ceiling connections (d), determine the location of the pulley. Of course, the actual value of $f(x)$, the distance the weight hangs below the ceiling, *does* depend on ℓ.

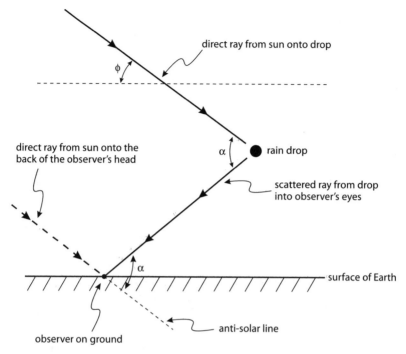

direct ray from sun onto drop

ϕ

direct ray from sun onto the
back of the observer's head

α ● rain drop

scattered ray from drop
into observer's eyes

α

surface of Earth

anti-solar line

observer on ground

FIGURE 5.8. Rainbows are sunlight scattered by raindrops.

5.8   Derivatives and the Rainbow

For the final section of this chapter, giving yet another illustration
of applying calculus to understand a physical problem, we return
to Snell's law. Imagine you are standing on a wide, level plain,
with the sun at your back, as shown in figure 5.8. The sun is angle
φ above the horizon. In front of you the sky is full of raindrops,
either because of a storm or, perhaps, because you are watering the
lawn with the garden hose set on spray. Some of the sun's light
rays will be scattered by the drops, i.e., through a combination of
internal reflections *and* refractions by the drops, light rays will be
bent backward and downward, into your eyes. This is the light you
see as the rainbow (more, later, on the colors), one of the most
beautiful of naturally occurring phenomena. As shown in figure 5.8,
if we extend the line from the sun to the observer (this is called the

anti-solar line) into the ground, then the *primary* rainbow appears at an angle α of about 42° up from the anti-solar line. (You'll soon see *why* that is so!) An immediate consequence of this, of course, is that there *is* a primary rainbow "there to see" only if the sun is sufficiently *low* in the sky so that the 42° "up angle" places the rainbow in the sky at all; this is clearly not the case if the sun is higher than 42° above the horizon. So, from the ground you can see rainbows in the morning, and in the afternoon, but never when the sun is directly overhead.

The search for understanding the origin of the rainbow was a long one, with speculations about it appearing in the writings of Aristotle (he incorrectly thought reflections off of *entire* clouds, rather than individual raindrops, was the mechanism involved). Indeed, when the first human eyes looked up into a passing rain shower, thousands of years ago, and saw the rainbow, who can doubt that awe wasn't inspired as much in primitive humans then as with Aristotle and us today? There often is also a secondary, much less bright rainbow (with the colors in reverse order) visible as well, at an angle of about 52° up from the anti-solar line. Why is that, and are there even more rainbows in the sky? People have wondered about such questions for centuries. The definitive history of the search for the answers is given in the book by the mathematician Carl Boyer, *The Rainbow: From Myth to Mathematics* (Princeton University Press 1987; first published in 1959). Two beautiful, nonmathematical books on the rainbow, each with many spectacular color images, are Robert Greenlear's *Rainbows, Halos, and Glories* (Cambridge University Press 1980; Professor Greenlear wrote the new introductory essay to the Princeton reprint of Boyer's book), and *Color and Light in Nature* (second edition) by David K. Lynch and William Livingston (Cambridge University Press 2001).

To start our analysis, we need to model *in detail* what happens to light rays arriving at a typical raindrop (assumed to be a sphere) in the sky. Figure 5.8 *is* pretty thin on detail! Figure 5.9 shows one such incident ray, and what happens to it.

1. As the ray arrives at point A on the drop's surface, a fraction of it is *reflected off* of the surface and the rest is *refracted into* the drop. The angles of incidence and refraction are, as in chapter 4, θ_i and θ_r, respectively.

2. The portion refracted into the drop travels onward until, upon arriving at point B on the back surface, a portion is *refracted* back out into space and the rest is internally reflected back into the drop. Since the sides *OA* and *OB* are equal in length (both are radii of the spherical drop), the triangle *OAB* is isosceles and the internal reflection angle is θ_r.

3. The internally reflected portion continues to "bounce around" inside the drop as it experiences a reflection/refraction with each additional interaction with the drop/air interface. Figure 5.9 shows only the refracted portion of the light ray that emerges from the drop at point *C* after the single reflection at point *B*. We'll come back later to the portions that go on to further adventures inside the drop, which will explain the secondary rainbow.

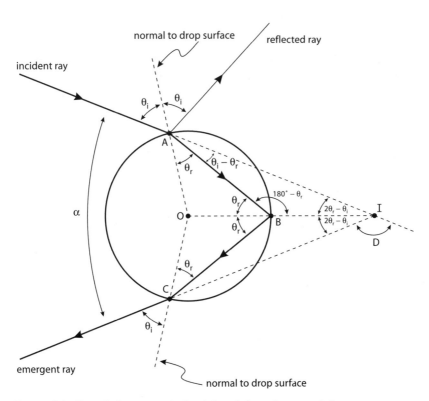

FIGURE 5.9. Detailed geometrical origin of the primary rainbow.

How *much* light is reflected and how much is refracted, at each interaction between the light ray and the water drop's air/water interface, is a complicated business. The answer depends on such details as the actual angle of incidence and the polarization of the light (which describes the electromagnetic details of the light). Fortunately, we don't have to go into the physics of light *that* far; all we care about here is that some light does emerge at the proper angle to arrive at our eyes. Light that goes elsewhere is light we don't see, in any case. For those who are interested in understanding how such intensity calculations are done, there is no better place to start than with the beautiful paper by Jearl D. Walker, "Multiple Rainbows from Single Drops of Water and Other Liquids" (*American Journal of Physics*, May 1976, pp. 421–33).

Concentrating for now on the ray emerging at point C, after just a single internal reflection (and two refractions), we extend (as shown in figure 5.9) the lines of the incident and the emergent rays until they intersect at point I. This defines the angle D, which is the total angular *deviation* suffered by the light ray from when it entered the drop until it left the drop. As shown by the geometry of figures 5.8 and 5.9, the angle (what we might call the *deflection* angle) between the incident and emerging rays is

$$\alpha = 2(2\theta_r - \theta_i),$$

and so

$$D = 180° - \alpha = 180° + 2\theta_i - 4\theta_r.$$

Now, from Snell's law we have, with n_1 and n_2, the indices of refraction of air and water, respectively:

$$\frac{\sin(\theta_i)}{\sin(\theta_r)} = \frac{n_2}{n_1} = n,$$

or

$$\theta_r = \sin^{-1}\left\{\frac{1}{n}\sin(\theta_i)\right\}$$

and

$$D = 180° + 2\theta_i - 4\sin^{-1}\left\{\frac{1}{n}\sin(\theta_i)\right\}.$$

In particular, if the incident ray passes through the center of the drop (through point O), then $\theta_i = 0°$ and thus the deflection angle is $\alpha = 0°$, and the deviation angle is $D = 180°$, i.e., the ray is reflected back out of the drop along the same path as it entered. The center-passing ray serves as an obvious reference axis.

Of course, the entire surface of the drop facing the sun receives rays, which we can assume are *parallel* rays because the sun is so very far away. There will be rays incident on the drop above the reference ray, and rays incident on the drop below the reference ray. From figure 5.9, which shows an incident ray above the center-passing reference ray, it is evident that such rays will emerge from *below* the reference ray. By symmetry, then, incident rays below the reference ray will emerge *above* the reference ray. An observer *on the ground* will therefore see rays of light emerging from the *bottom* portion of a drop due to incident rays illuminating the *upper* portion of the drop.

The observed rays come out of the drop with various values for the angle α, but not *all* values are equally likely. This is easy to see if we simply plot α as a function of where the incident ray strikes the drop. The geometry of this is illustrated in figure 5.10, which shows the center-passing ray as the *horizontal* (for ease in drawing) reference axis, and a typical ray incident on the drop *above* the reference ray. We can calculate the angle of incidence, θ_i, as

$$\sin(\theta_i) = \frac{y}{R}, \qquad 0 \le y \le R,$$

where R is the radius of the spherical drop and y is the vertical displacement of the incident ray from the reference ray. The reason for formulating the mathematics this way is because, for a drop in uniform sunlight, there are no preferred values for y. That is, in loose probabilistic jargon, of all the rays striking the drop, a randomly selected ray is as likely to have one value of y as any other (this is

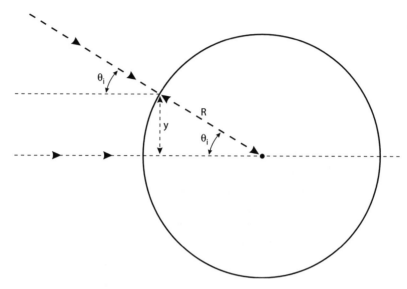

FIGURE 5.10. Illuminating a raindrop.

not true for θ_i; θ_i is *not* uniformly distributed from 0° to 90° for a spherical drop in uniform sunlight).

From Snell's law, we have

$$\theta_r = \sin^{-1}\left\{\frac{1}{n}\sin(\theta_i)\right\} = \sin^{-1}\left\{\frac{1}{n}\cdot\frac{y}{R}\right\},$$

and so

$$\alpha = 4\theta_r - 2\theta_i = 4\sin^{-1}\left(\frac{y}{nR}\right) - 2\sin^{-1}\left(\frac{y}{R}\right), \qquad 0 \le y \le R.$$

Figure 5.11 shows a plot of the angle α as y/R varies from 0 to 1 (the actual value of R is, then, for our simple analysis here, unimportant), using the value of $n = 4/3$ for a water drop in air, and it is obvious that α has a maximum at about 42°. (Aha!—the rainbow angle I mentioned at the start of this section. You are *almost* at the point of understanding the *physical* significance of this angle.) This is interesting, yes, but what makes it *really* interesting is that it is a *broad* maximum, i.e., there is a concentration of rays with α-angles at and around 42°. For example, 20% of the emergent light

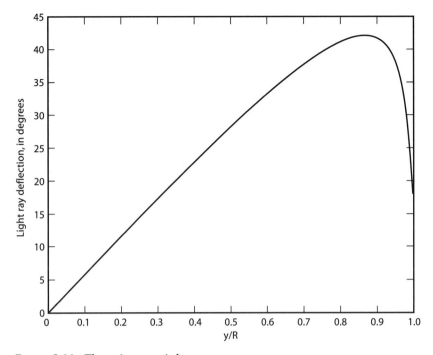

FIGURE 5.11. The primary rainbow.

(0.75 $\leq$ y/R $\leq$ 0.95) has an α-angle in the narrow interval from 40° to 42°. The other 80% of the emergent light is (more or less) uniformly distributed over the much larger α-angle interval of 0° to 40°.

We can calculate the precise value of the maximum α directly, using calculus. Since $\alpha = 2(2\theta_r - \theta_i) = 4\theta_r - 2\theta_i$, then

$$\frac{d\alpha}{d\theta_i} = 4\frac{d\theta_r}{d\theta_i} - 2$$

and, setting the derivative equal to zero at the maximum of α gives

$$\frac{d\theta_r}{d\theta_i} = \frac{1}{2}.$$

Then, differentiating Snell's law (using the chain rule) with respect to θ_i, i.e., differentiating $\sin(\theta_i) = n\sin(\theta_r)$, we get

$$\cos(\theta_i) = n\cos(\theta_r)\frac{d\theta_r}{d\theta_i} = \frac{1}{2}n\cos(\theta_r).$$

Thus,

$$\cos^2(\theta_i) = \frac{1}{4}n^2\cos^2(\theta_r),$$

and since $\cos^2(\theta_r) = 1 - \sin^2(\theta_r)$, then

$$\cos^2(\theta_i) = \frac{1}{4}n^2\left[1 - \sin^2(\theta_r)\right] = \frac{1}{4}n^2\left[1 - \frac{1}{n^2}\sin^2(\theta_i)\right],$$

or

$$4\cos^2(\theta_i) = n^2 - \sin^2(\theta_i).$$

Since $\sin^2(\theta_i) = 1 - \cos^2(\theta_i)$, this becomes

$$4\cos^2(\theta_i) = n^2 - 1 + \cos^2(\theta_i),$$

or, when α is maximum, θ_i is given by

$$\theta_i = \cos^{-1}\left\{\sqrt{\frac{n^2-1}{3}}\right\}.$$

Now, as before,

$$\alpha = 4\theta_r - 2\theta_i = 4\sin^{-1}\left[\frac{1}{n}\sin(\theta_i)\right] - 2\theta_i,$$

and so, finally,

$$\alpha_{max} = 4\sin^{-1}\left[\frac{1}{n}\sin\left\{\cos^{-1}\left(\sqrt{\frac{n^2-1}{3}}\right)\right\}\right] - 2\cos^{-1}\left\{\sqrt{\frac{n^2-1}{3}}\right\}.$$

For $n = 4/3$, this reduces to

$$\alpha_{max} = 4\sin^{-1}\left[\frac{3}{4}\sin\left[\cos^{-1}\left(\frac{1}{3}\sqrt{\frac{7}{3}}\right)\right]\right] - 2\cos^{-1}\left\{\frac{1}{3}\sqrt{\frac{7}{3}}\right\}$$

$$= 42.03°,$$

just as shown in figure 5.11, and as had been known by direct observation of rainbows for centuries *before* Descartes.

The crucial observation, of a *broad* maximum for α, *is* due to Descartes, who discovered the concentration of emergent rays at $\alpha = 42°$ by a laborious tracing of many incident rays (using Snell's law) through a single gigantic, artificial drop (a glass spherical globe). As he wrote in *Les Météores* ("Meteorology"), another appendix to his 1637 *Discours de la Méthode*:

> I took my pen and calculated in detail all the rays which fall on the various points of a drop of water, in order to see under what angles they could come toward our eyes after two refractions and one or two reflections. I found that after one reflection and two refractions, very many more of them can be seen under the angle of 41° to 42° than under any lesser one; and that none of them can be seen under a larger angle. [This gives the primary rainbow.] Then I also found that after two reflections and two refractions, very many more of them come toward the eye under a 51° to 52° angle, than under any larger one; and no such rays come under a lesser [angle]. [This gives the secondary rainbow, as you'll see soon.]

The concentration of light rays around the extrema of α is, of course, in the very nature of an extrema; i.e., rays with α-angles on either side of α_{max} have nearly equal α-angles. We can now understand the first central question about rainbows—why do they appear as *circular* arcs in the sky? Geometry, alone, answers that. Figure 5.8 shows just a single raindrop scattering a ray of light into the eyes of an observer on the ground. That drop, the observer, and the parallel rays of light incident on the drop, are all shown in the same plane (the plane defined by the page the figure is printed on). In the actual world, however, there are infinitely many planes that contain the observer, raindrops, and parallel light rays incident on those raindrops. Those drops also reflect light rays into the eyes of the observer because, as a little mental imagery should convince you, all of the geometry of figures 5.8 and 5.9 are preserved if we *rotate* those figures around the anti-solar line. That is, all of the raindrops scattering light back into the observer's eyes lie on the surface of a cone with a central angle of about 84° (angular *radius* of 42°) and the anti-solar line as its axis; the light entering the observer's eyes appears to come from a *circular arc*.

But notice, too, that the distance of any particular drop from the observer's eyes is not important; *all* of the drops on the cone's surface scatter light rays back down to the observer's eyes, and so the rainbow is not in any particular place in the sky. The drops can be mere inches away, as well as many miles distant. And notice, too, that "the" cone is different for different observers. That means each observer receives scattered light from different sets of raindrops and so, while different observers see a rainbow, it is not the *same* rainbow. Indeed, each eye of a *lone* observer "sees" a different rainbow. The rainbow, then, as befits its ethereal beauty, is literally nowhere in particular and everywhere in general; if it is anywhere, it is in "the eye of the beholder"!

The second central question—why is the rainbow multicolored? —requires more than geometry, which is why the answer escaped Descartes and all those before him. The explanation comes from the fact that n is not a constant; the value $n = 4/3 (= 1.333)$ I used to compute α_{max} is simply a typical value of the refractive index of water in the visible portion of the spectrum. As mentioned at the end of the previous chapter, n depends on the frequency (color) of the light rays, with $n = 1.344$ for violet and $n = 1.331$ for red (the extreme ends of the visible spectrum). There will therefore be a different value for α_{max} for each color, and the various colors will appear as distinctly separate but adjacent rainbows. If you run the red and violet values for n through the equation for α_{max}, the numbers work out to be

$$\alpha_{max} \text{ (for red)} = 42.37°$$

$$\alpha_{max} \text{ (for violet)} = 40.5°.$$

Since $\alpha_{max}(\text{red}) > \alpha_{max}(\text{violet})$, the red rainbow appears higher in the sky than does the violet rainbow, and so red and violet are predicted to define the outer and inner edge colors of the rainbow, respectively—just as is observed.

Now, what of those light rays inside the drop shown in figure 5.9 that do *not* exit the drop after just one internal reflection, but rather after *two* such reflections, and then enter the eyes of an observer on the ground? Such a light ray is shown in figure 5.12, which illustrates the fact that, for the exiting ray to be directed downward to earth, the incident ray must arrive at the bottom portion of

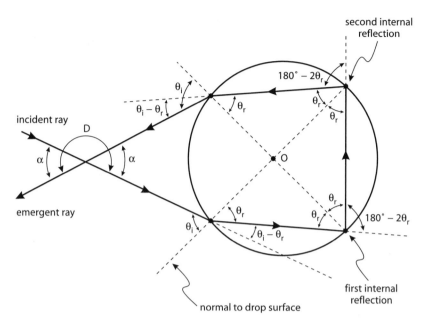

FIGURE 5.12. Detailed geometrical origin of the secondary rainbow.

the drop and emerge from the upper portion. This is precisely the opposite of what is depicted in figure 5.9, which gives rise to the primary rainbow. The situation shown in figure 5.12 will give us, instead, the *secondary* rainbow. Figure 5.12 again defines the angles α and D as, respectively, the angle between the incident and emergent rays, and the total angular deviation experienced by the light ray from when it enters the drop until it leaves the drop. From figure 5.12 it is clear that now $D = 180° + \alpha$, i.e.,

$$\alpha = D - 180°,$$

whereas in figure 5.9 (for a single internal reflection) we had $\alpha = 180° - D$.

To find α, which is the "up-angle" from the anti-solar line to the (secondary) rainbow, I'll first find D and then subtract 180°. From the geometry of figure 5.12, we see that when the incident ray enters the drop by refraction it suffers an initial deviation of $\theta_i - \theta_r$, and then at *each* internal reflection it undergoes an additional deviation of $180° - 2\theta_r$. Finally, at the second refraction that produces the emergent ray, there is a final deviation, again, of $\theta_i - \theta_r$. (Indeed, a look back at figure 5.9 shows we could have calculated D for the primary rainbow in this manner rather than the way actually used.) Thus, with two internal reflections, we have

$$D = (\theta_i - \theta_r) + 2(180° - 2\theta_r) + (\theta_i - \theta_r)$$
$$= 360° + 2\theta_i - 6\theta_r.$$

And so

$$\alpha = D - 180° = 180° + 2\theta_i - 6\theta_r,$$

or

$$\alpha = 180° - 2(3\theta_r - \theta_i).$$

Now, just as we did before, let's imagine a drop in uniform sunlight and plot α as a function of y/R, where y is (again) the vertical displacement of the incident ray from a horizontal center-passing reference ray, and R is the radius of the drop. And, as before, from Snell's law we have

$$\theta_r = \sin^{-1}\left\{\frac{1}{n} \cdot \frac{y}{R}\right\}$$
$$\theta_i = \sin^{-1}\left(\frac{y}{R}\right),$$

and so

$$\alpha = 180° - 2\left[3\sin^{-1}\left\{\frac{1}{n} \cdot \frac{y}{R}\right\} - \sin^{-1}\left\{\frac{y}{R}\right\}\right], \qquad 0 \le \frac{y}{R} \le 1.$$

Figure 5.13 shows the result for $n = 4/3$, with α now exhibiting a *minimum* (rather than the maximum we got for the primary rainbow) at about 52°, just as reported by Descartes. Thus, we have the secondary rainbow at about 10° higher in the sky than the primary rainbow, which is just as observed when the secondary rainbow can, in fact, be observed (it is, of course, much less bright than the primary—only about 43% as bright—because of the additional loss

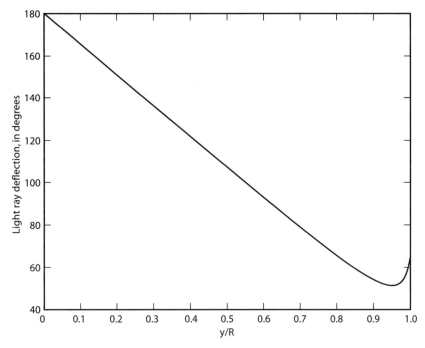

FIGURE 5.13. The secondary rainbow.

of intensity from the further adventures the light rays experience within the water drops). Again, if we examine α as a function of n (color), we find that α_{min} is different for different colors, but now there is (literally) a new twist—the color sequence in the secondary rainbow is the reverse of the primary rainbow. That is, the red rainbow will appear lower in the sky than does the violet rainbow, and so red is predicted to be the *inner* edge color (and violet the *outer* edge color) of the secondary rainbow. And that is precisely what is seen.

The secondary rainbow has occasionally appeared in fictional literature. In her episodic novel *Strange Attractors* (Viking 1993), for example, Rebecca Goldstein ends her story with a description of a group of mathematicians running outdoors to observe a double rainbow. Her words are lovely to read, but flawed ever so slightly by positioning the secondary rainbow in the wrong part of the sky:

> And outside the mathematicians all stand gathered together on the wet lawn, staring up into the western sky, where there's a rare double rainbow stretching itself: The colors of the primary arc are intense and *beneath* [my emphasis] is the secondary rainbow, with its paler inversion of the spectrum. And all of the mathematicians are standing together in silence; on every face the same look of transfixed bliss.

As was Descartes' practice, he failed (for whatever reason) to acknowledge the prior work of others into the nature of the rainbow. In fact, the 42° angle of the primary rainbow can be found in the *Opus Majus* (1267) of the English philosopher and Franciscan friar Roger Bacon. And it was only a few decades later, in 1304, that the German monk Dietrich von Freiberg (1250–1310) advanced the correct explanation for the rainbow as the scattering of light by *individual* raindrops. It is in his writing, too, that we find the conclusion that each observer sees a personal rainbow from different sets of drops. And not only that, it was Theodoric of Freiberg (as he is generally called in the English literature) who was the first to experiment with water-filled transparent containers, as artificial drops, to trace the paths of light rays. And not only *that*, it was Theodoric of Freiberg, not Descartes, who was the first to associate the primary rainbow with two refractions and one internal reflection, and the secondary rainbow with two internal refractions and two reflections. His small book *De iride et radialibus impressionibus* ("On the Rainbow and 'Radiant Impressions'") put forth all of these fundamental ideas, but not a word about any of it appears in *Les Météores*. To give Descartes his due, however (which is more than he did for *his* predecessors), discovery of the *concentration* of rays at the observed rainbow angle is Descartes' alone.

The primary (secondary) rainbow is the result of two refractions and one (two) internal reflection(s). Wouldn't three internal reflections therefore produce yet another rainbow (the so-called *tertiary* rainbow)? And why stop there—what of the possibility of rainbows produced after N internal reflections, where N is *any* positive integer? Such high-order rainbows would, of course, be expected to be even dimmer, but perhaps sufficiently sensitive eyes could see them—if they exist. The question of higher-order rainbows, particularly the tertiary, intrigued many people over the centuries, and they searched the sky for them. The logical place to look would seem to be in the sky above the secondary, which itself is 10° above the primary. Despite occasional claims to have seen the third-order rainbow, however, nobody ever *has* seen it, and nobody ever will—even though it surely *does* exist! Calculus explains this apparent paradox, with a calculation first done by Newton, probably some time around 1670.

An easy extension of the analysis just done for D, the angle of total deviation experienced by a light ray from when it first arrives on the surface of a drop until it exits the drop, leads to the result

$$D = (\theta_i - \theta_r) + N(180° - 2\theta_r) + (\theta_i - \theta_r)$$
$$= 2(\theta_i - \theta_r) + N(180° - 2\theta_r)$$

if there are N internal reflections (for the particular case of the secondary rainbow, we used $N = 2$). Notice that for $N = 1$, the case of the primary rainbow, this expression reduces to $D = 180° + 2\theta_i - 4\theta_r$, which is, indeed, the result arrived at in the discussion at the start of this section. As was the case for the first two rainbows, *all* rainbows occur at the extrema of the angle D. So, differentiation of D with respect to θ_i gives

$$\frac{dD}{d\theta_i} = 2 - 2\frac{d\theta_r}{d\theta_i} - 2N\frac{d\theta_r}{d\theta_i},$$

which, when set equal to zero, says

$$\frac{d\theta_r}{d\theta_i} = \frac{1}{1 + N}$$

when D (for the Nth order rainbow) is at its extrema (which is what gives rise to the *concentration* of observed light rays, i.e., the rainbow).

Remembering the result we calculated earlier from a differentiation of Snell's law,

$$\cos(\theta_i) = n \cos(\theta_r)\frac{d\theta_r}{d\theta_i},$$

we therefore have

$$\frac{d\theta_r}{d\theta_i} = \frac{1}{n} \cdot \frac{\cos(\theta_i)}{\cos(\theta_r)} = \frac{1}{1+N}.$$

Cross-multiplication and squaring gives

$$(N+1)^2 \cos^2(\theta_i) = n^2 \cos^2(\theta_r).$$

From trigonometry and Snell's law, we have

$$\cos^2(\theta_r) = 1 - \sin^2(\theta_r) = 1 - \frac{1}{n^2}\sin^2(\theta_i),$$

and so

$$(N+1)^2 \cos^2(\theta_i) = n^2 - \sin^2(\theta_i) = n^2 - \left[1 - \cos^2(\theta_i)\right],$$

or

$$(N+1)^2 \cos^2(\theta_i) = n^2 - 1 + \cos^2(\theta_i).$$

This is easily solved for $\cos(\theta_i)$ to give Newton's equation for the condition that must be satisfied for the Nth order rainbow:

$$\boxed{\cos(\theta_i) = \sqrt{\frac{n^2 - 1}{N(N + 2)}}.}$$

As a check, notice that for the case of $N = 1$ (the primary rainbow) this does reduce correctly to the result calculated earlier: $\cos(\theta_i) = \sqrt{(n^2 - 1)/3}$.

So, *where is* the tertiary rainbow? Inserting $N = 3$ (and using $n = 4/3$), we have

$$\cos(\theta_i) = \sqrt{\frac{\left(\frac{4}{3}\right)^2 - 1}{(3)(5)}} = \frac{1}{3}\sqrt{\frac{7}{15}},$$

or $\theta_i = 76.84°$. And so, from Snell's law,

$$\sin(\theta_r) = \frac{1}{n}\sin(\theta_i) = \frac{3}{4}\sin(76.84°),$$

or

$$\theta_r = \sin^{-1}\left\{\frac{3}{4}\sin(76.84°)\right\} = 46.91°.$$

Thus,

$$D = 2(76.84° - 46.91°) + 3[180° - 2(46.91°)]$$

$$= 318.4°.$$

To understand what this value means, take a look at figure 5.14, which shows the case of $N = 3$ internal reflections. It is clear from the geometry there that the ray enters through the bottom portion

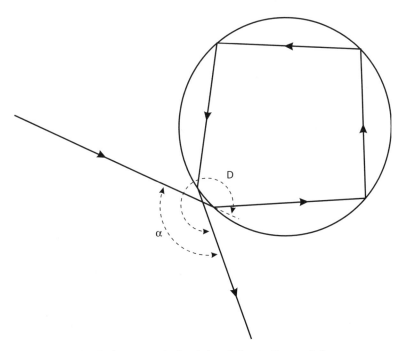

FIGURE 5.14. Detailed geometrical origin of the tertiary rainbow.

of the drop, bounces once nearly all around the inside of the drop, exits the bottom portion of the drop, and so is scattered downward and *forward* out of the drop. That is, for an observer on the ground to see the scattered ray, she must *turn around* and look behind her! The tertiary rainbow is indeed "there" (*if* there are raindrops in the skies between the sun and the observer), and it is indeed higher in the sky than is the secondary. It is actually higher than straight up. Until Newton's calculation, people had been looking forward with the sun behind them, just above the secondary rainbow, and that's simply the *wrong* place to look. But even if somebody had turned around, they still wouldn't have seen the tertiary rainbow because, in addition to its inherent dimness (the tertiary is only about 24% as bright as the primary), it is completely overwhelmed by the nearby glare of the sun, as shown in the following box. And that's why nobody ever will see the *natural* tertiary rainbow. Artificially produced rainbows, generated in the laboratory with a laser playing the role of the sun, have let experimenters actually see rainbows up to at least $N = 20$. They are right where Newton's boxed equation for $\cos(\theta_i)$ says they should be.

The tertiary rainbow has one last surprise for us—its shape. Celebrity intellectual Marilyn vos Savant stumbled on this point when replying to a reader's question on where the third-order rainbow is. In her *Parade Magazine* column "Ask Marilyn" (August 4, 2002), she wrote that the tertiary rainbow "*arches over* [my emphasis] the second [i.e., secondary] one." This is not so, and here's why.

Just as the primary and secondary rainbows are rotationally symmetric around the *anti*-solar line (the observer is facing *away* from the sun), the tertiary rainbow is rotationally symmetric around the *solar* line (the line from the sun to the observer who is now facing the sun). As shown in figure 5.15, the scattered ray from a typical raindrop that is between the sun and the observer makes an angle of (about) 41.6° with respect to the solar line. Because of the rotational symmetry, then, scattered light rays reach the observer from raindrops on the surface of a cone (with the solar line as the axis) with an

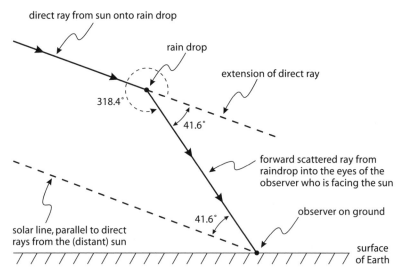

direct ray from sun onto rain drop

rain drop

extension of direct ray

318.4°

41.6°

forward scattered ray from
raindrop into the eyes of the
observer who is facing the sun

observer on ground

41.6°

solar line, parallel to direct
rays from the (distant) sun

surface
of Earth

FIGURE 5.15. Locating the tertiary rainbow in the sky.

angular radius of 41.6°. That is, the tertiary rainbow is a *circular
halo around the sun*!

The "discovery" of the tertiary's place in the sky has an interesting history. After Newton was appointed in late 1669, at age 26, to
the Lucasian Professorship of Mathematics at Cambridge (the chair
now held by the famous theoretical physicist Stephen Hawking), he
gave a series of inaugural lectures during the period 1670–72. Those
lectures were not published at the time (they are available to the
modern reader in *The Optical Papers of Isaac Newton*, edited by Alan
E. Shapiro, Cambridge University Press 1984), but they did serve as
the basis for his 1704 book *Opticks*. In his optical lectures, Newton
discussed the rainbow, including calculations concerning rainbows
beyond the secondary. It isn't entirely clear if he used his general
results to actually calculate the specific angular position of the tertiary rainbow (he certainly didn't publish it), and in *Opticks* he wrote
only that "The light which passes through a drop of rain after two
refractions, and three or more reflections, is scarcely strong enough
to cause a sensible bow." There is no mention of the glare of the
sun overwhelming the tertiary rainbow halo. Johann Bernoulli later

reproduced Newton's general approach, but he too failed to specifically calculate the location of the tertiary. Like Newton, Bernoulli made only a single suggestive comment, to the effect that while the tertiary rainbow might be visible to eagles or lynxes, it would not be visible to human eyes. Where in the heavens is home to the tertiary rainbow was finally specifically published in 1700, in the *Philosophical Transactions of the Royal Society*. The author was Edmond Halley, Newton's friend and the cannon-shooter extraordinaire from earlier in this chapter.

And, finally, to end this chapter on a cosmological note, what will rainbows look like in the very far future? This question is not quite as odd as you might think—when the sun is vastly older than it is now, it will be much less hot, and its spectrum will be predominantly at the longer (infrared) wavelengths. Will there still be a rainbow to be "seen"? There will be, indeed, but only seen by (nonhuman?) eyes that have adapted to the shifted spectrum. We know this is so, because there is an infrared rainbow in the sky *right now*, and it has been photographed. You can read how that was done in the article by Robert Greenlear, "Infrared Rainbow" (*Science*, September 24, 1971, pp. 1231–32). Professor Greenlear ends on a poetic note, writing of his "fascination in 'seeing' for the first time an infrared rainbow which has hung in the sky undetected since before the presence of man on this planet." And he found it right where all the math theory of this chapter says it should be.

Solution to Steiner's Problem in Section 5.1

With the Steiner function written as

$$f(x) = x^{\frac{1}{x}} = e^{\ln\left(x^{1/x}\right)} = e^{\frac{1}{x}\ln(x)} = e^{g(x)},$$

we have

$$g(x) = \frac{1}{x}\ln(x).$$

Now, from our result in section 4.5 on how to differentiate a composite function,

(continued)

$$\frac{df}{dx} = \left\{\frac{d}{dg}e^g\right\} \cdot \left\{\frac{dg}{dx}\right\} = e^g\left[\frac{1}{x^2} - \frac{1}{x^2}\ln(x)\right],$$

or

$$\frac{df}{dx} = \frac{x^{\frac{1}{x}}}{x^2}[1 - \ln(x)].$$

Since $x^{1/x}/x^2 > 0$ for all $x > 0$, then $f'(x) = 0$ only when $1 - \ln(x) = 0$, i.e., when $x = e$.

We could now use the second derivative test to show that $x = e$ gives a maximum, but we can see this more directly by simply observing that

$$f(1) = 1$$
$$f(2) = 2^{1/2} > f(1)$$
$$f(3) = 3^{1/3} > f(2)$$
$$f(4) = 4^{1/4} = (2^2)^{1/4} = 2^{1/2} = f(2) < f(3).$$

That is,

$$f(1) < f(2) < f(3) > f(4),$$

and so we expect $f(x)$ to have a *maximum* at some x between 2 and 4; notice that $e = 2.718\ldots$ satisfies that requirement. (Do you see *why* $f(3) = 3^{1/3} > f(2) = 2^{1/2}$? Just raise both quantities to the sixth power and observe that $f^6(3) = 3^2 = 9$ and $f^6(2) = 2^3 = 8$. Since $9 > 8$, then $f(3) > f(2)$.) Thus, the maximum value of $f(x)$ is

$$f(e) = e^{1/e} = 1.444667861\ldots,$$

often called *Steiner's number*.

6.

Beyond Calculus

6.1 Galileo's Problem

The story of Galileo Galilei (1564–1642), and of his research into the physics of free-fall by dropping various weights from the top of the Leaning Tower of Pisa, is too well known to be retold here. Whatever the truth of the details of that story, it is undeniable that the Italian astronomer was deeply interested in how things move under the influence of gravity. It was that interest that eventually led to what is generally thought to be the first solved problem in the calculus of variations, which was the next great step beyond the calculus of Newton and Leibniz in solving minimization problems. Galileo's own attempt at the original version of the problem was one of mixed success and, indeed, one that still prompts some debate among historians.

Galileo did the work that set the stage for the ultimate version of the so-called "minimum descent time" problem during the final, most troubled years of his life, troubles caused by his belief in Copernicanism. Copernicanism teaches that all the planets (including Earth) orbit the sun, not a stationary Earth. In direct contradiction with Biblical scripture, such a belief was bound to lead to a collision with the Church. After the 1632 publication of his book *Dialogue Concerning the Two Chief World Systems*, in which he advocated positions in conflict with religious teachings, Galileo was summoned to Rome in 1633 on the charge of suspicion of heresy. He was sick in bed when summoned, and so he declined to make the journey. He perhaps first realized how precarious was his position when the Pope (a friend of many years!) threatened to forcibly transfer him to Rome, in chains, if he continued to refuse. So he went, but the "trial" was a farce, with an outcome no one could doubt.

His very life hung in the balance, and he was lucky to get off with "only" the placing of the *Dialogue* on the Index (of forbidden books), a prohibition against ever publishing again, being forced to recant, and imprisonment (later commuted to house arrest, with surveillance, for life). Although now very sick and nearly blind, Galileo proved to be tougher than the religious thugs of the Inquisition; he used his cruel confinement to write one more book, *Discourses and Mathematical Demonstrations Concerning Two New Sciences*. It was smuggled out of Italy and published in Holland in 1638, just as Descartes and Fermat were doing battle in France over Snell's law. Galileo's new, groundbreaking ideas on how things fall in gravity were described in that final work.

Imagine a bead with a wire threaded through a hole in it, such that the bead can slide (with no friction) along the wire. Suppose that the wire is bent into the shape of a circular arc with radius L, and positioned vertically. The bead is held at point D, as shown in figure 6.1, so that the radius to the bead makes angle α with the

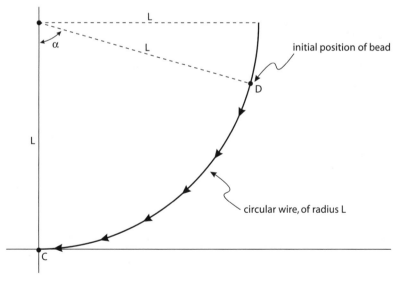

FIGURE 6.1. A bead sliding under gravity along a vertical, circular wire.

vertical. We then release the bead, which slides to the bottom of the wire at point C. That is, the bead makes a *circular descent* under the influence of gravity. An immediate and natural question to ask is, how long does the descent take? It was far beyond the mathematics of Galileo's day to compute the precise answer, and his approach to the problem is via ingenious geometrical constructions. Today we *can* compute the answer (see appendix E for the details): if T is the descent time, and g denotes the acceleration of gravity, then

$$
T = \sqrt{\frac{L}{g}} \int_0^{\frac{\pi}{2}} \frac{d\beta}{\sqrt{1 - k^2 \sin^2(\beta)}}, \qquad k = \sin\left(\frac{1}{2}\alpha\right),
$$

an expression that would have been meaningless to Galileo (or, for that matter, to any other mathematician of the first half of the seventeenth century). Instead, Galileo used inclined planes as approximations to a circular arc to calculate approximations to the time of descent. What I'll show you here is a modern treatment of Galileo's ideas, although his development was strictly geometric (and *very* subtle). You can find the original geometric approach discussed in the paper by Herman Erlichson, "Galileo's Work on Swiftest Descent from a Circle and How He Almost Proved the Circle Itself Was the Minimum Time Path" (*American Mathematical Monthly*, April 1998, pp. 338–47).

As Professor Erlichson pointed out in an earlier paper ["Galileo's Pendulums and Planes" (*Annals of Science*, May 1994, pp. 263–72)], the original motivation for Galileo's interest in the question of the descent time along a vertical circular path came from his interest in pendulum motion; a light fixture hanging from a chain attached to the ceiling of a church executes a circular swing when disturbed by an earthquake. The *period* of such a swing (the time for one complete swing, from the starting point of the fixture back to the starting point) would thus be given by $4T$, a value Galileo incorrectly believed to be independent of α (the amplitude of the swing). Galileo was wrong but, actually, not by very much.

The first, crudest approximation to circular descent, using straight line segments (or *inclined planes*, as Galileo thought of the approximations), would be descent along the direct, single segment connecting D and C. The next, somewhat less crude approximation

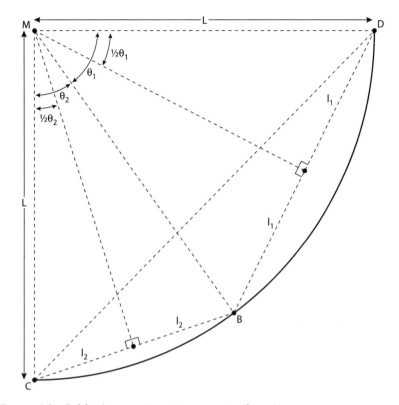

FIGURE 6.2. Galileo's approximation to a circular wire.

would use the broken line descent along *two* inclined planes (D to B, then B to C), as shown in figure 6.2. The arc DBC is, as drawn in that figure, one-quarter of a circle of radius L, centered on M. Point B is arbitrary, with the radius from M to B making angle θ_1 with the radius from M to D (if $\theta_1 = 0°$ then $B = D$, and if $\theta_1 = 90°$ then $B = C$). What I'll do next is derive T_D and T_B, the times for the bead (starting from rest) to slide under gravity from D to C along the *Direct* path and along the *Broken* path, respectively.

The calculation of T_D is easy, once you observe that the bead's speed during the descent increases *linearly* from $v_D = 0$ at D to $v_C = \sqrt{2gL}$ at C. The linear part follows from the fact that, along the entire, direct path from D to C, the acceleration of the bead by gravity is constant. The expression for v_C follows from simply equating the change in the bead's kinetic energy of motion from

D to C to the change in its potential energy of position (since we are assuming zero friction, then conservation of energy holds). So, if the bead has mass m,

$$\frac{1}{2}mv_C^2 = mgL,$$

and so, as claimed,

$$v_C = \sqrt{2gL}.$$

The average speed of the descent is then given by

$$\frac{v_C + v_D}{2} = \frac{1}{2}\sqrt{2gL}.$$

The length of the direct path from D to C is obviously

$$\sqrt{L^2 + L^2} = L\sqrt{2},$$

and so

$$T_D = \frac{L\sqrt{2}}{\frac{1}{2}\sqrt{2gL}} = 2\sqrt{\frac{L}{g}}.$$

As shown in appendix E, if $\alpha = 90°$, then the time for true circular descent on the quarter circle is $T = 1.8541\sqrt{L/g}$, and so T_D is less than 8% longer than T, i.e.,

$$\frac{T_D}{T} = \frac{2\sqrt{\frac{L}{g}}}{1.8541\sqrt{\frac{L}{g}}} = 1.0787.$$

Galileo didn't know this, but he *did* know one astonishing fact about T_D—it is independent of the position of D. In the above discussion, I took D as at the top end of a quarter-circular arc, but if we instead let the radius from M to D be at some (arbitrary) angle θ below the full quarter circle (see figure 6.3) we'll find the descent time remains

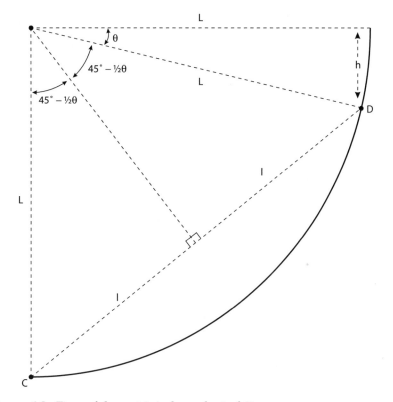

F<small>IGURE</small> 6.3. Time of descent is independent of D.

unchanged. This is, I think, not at all obvious, but it is not hard to demonstrate.

Since we now have the bead's initial position, D, decreased vertically by the amount $h = L \sin(\theta)$, then the vertical drop of the bead during its descent is $L - L \sin(\theta)$. Thus, its speed at C is

$$v_C = \sqrt{2gL\{1 - \sin(\theta)\}}$$

and its average speed during the descent is $\frac{1}{2}v_C$, just as before. The length of the direct path is now

$$2\ell = 2L \sin\left(45° - \frac{1}{2}\theta\right),$$

and so the time of descent (T_θ) is (in our notation, $T_D = 2\sqrt{L/g}$ is the special case of $T_{\theta=0°}$)

$$T_\theta = \frac{2L \sin\left(45° - \frac{1}{2}\theta\right)}{\frac{1}{2}\sqrt{2gL\{1 - \sin(\theta)\}}} = 2\sqrt{\frac{L}{g}}\sqrt{2}\frac{\sin\left(45° - \frac{1}{2}\theta\right)}{\sqrt{1 - \sin(\theta)}},$$

or

$$T_\theta = T_D \left\{ \sqrt{2}\frac{\sin\left(45° - \frac{1}{2}\theta\right)}{\sqrt{1 - \sin(\theta)}} \right\}.$$

This looks complicated but, in fact, the quantity in the braces equals one *for all* θ! This is so because, from the trigonometric addition identity for the sine,

$$\sqrt{2}\frac{\sin\left(45° - \frac{1}{2}\theta\right)}{\sqrt{1 - \sin(\theta)}} = \sqrt{2}\frac{\sin(45°)\cos\left(\frac{1}{2}\theta\right) - \cos(45°)\sin\left(\frac{1}{2}\theta\right)}{\sqrt{1 - \sin(\theta)}}$$

$$= \sqrt{2}\frac{\frac{1}{\sqrt{2}}\cos\left(\frac{1}{2}\theta\right) - \frac{1}{\sqrt{2}}\sin\left(\frac{1}{2}\theta\right)}{\sqrt{1 - \sin(\theta)}} = \frac{\cos\left(\frac{1}{2}\theta\right) - \sin\left(\frac{1}{2}\theta\right)}{\sqrt{1 - \sin(\theta)}}.$$

If we square this last expression and then use the trigonometric identity $\sin(\alpha)\cos(\beta) = \frac{1}{2}\{\sin(\alpha - \beta) + \sin(\alpha + \beta)\}$, we get

$$\frac{\cos^2\left(\frac{1}{2}\theta\right) - 2\cos\left(\frac{1}{2}\theta\right)\sin\left(\frac{1}{2}\theta\right) + \sin^2\left(\frac{1}{2}\theta\right)}{1 - \sin(\theta)}$$

$$= \frac{1 - 2 \cdot \frac{1}{2}\left\{\sin\left(\frac{1}{2}\theta - \frac{1}{2}\theta\right) + \sin\left(\frac{1}{2}\theta + \frac{1}{2}\theta\right)\right\}}{1 - \sin(\theta)} = \frac{1 - \sin(\theta)}{1 - \sin(\theta)} = 1,$$

and so $T_\theta = T_D$, for *any* θ, not just for $\theta = 0°$. Amazing!

Let's next calculate T_B, the descent time along the broken path DBC in figure 6.2. As before, $v_D = 0$ at D. To get to B, the bead falls through a vertical distance of $L \sin(\theta_1)$ and so $v_B = \sqrt{2gL \sin(\theta_1)}$. And, as before, at the end of the descent, $v_C = \sqrt{2gL}$. Also as before, since the accelerations on DB and BC are constant (although *not* equal), then the speed of the bead along each segment increases linearly from its initial speed to its final speed on each segment. So, the average speed on DB is $\frac{1}{2}\sqrt{2gL \sin(\theta_1)}$, and the average speed on BC is $\frac{1}{2}\{\sqrt{2gL} + \sqrt{2gL \sin(\theta_1)}\}$. The lengths of the two segments are $\overline{DB} = 2\ell_1 = 2L \sin(\frac{1}{2}\theta_1)$ and $\overline{BC} = 2\ell_2 = 2L \sin(\frac{1}{2}\theta_2)$ and thus, the time of descent, along the two-segment broken path from D to C, is

$$
T_B = \frac{2L \sin\left(\dfrac{1}{2}\theta_1\right)}{\frac{1}{2}\sqrt{2gL \sin(\theta_1)}} + \frac{2L \sin\left(\dfrac{1}{2}\theta_2\right)}{\frac{1}{2}\{\sqrt{2gL} + \sqrt{2gL \sin(\theta_1)}\}}
$$

$$
= 2\sqrt{\frac{2L}{g}} \cdot \frac{\sin\left(\dfrac{1}{2}\theta_1\right)}{\sqrt{\sin(\theta_1)}} + 2\frac{2L \sin\left(\dfrac{1}{2}\theta_2\right)}{\sqrt{2gL}\{1 + \sqrt{\sin(\theta_1)}\}}
$$

$$
= 2\sqrt{2}\sqrt{\frac{L}{g}} \cdot \frac{\sin\left(\dfrac{1}{2}\theta_1\right)}{\sqrt{\sin(\theta_1)}} + 2\sqrt{2}\sqrt{\frac{L}{g}} \cdot \frac{\sin\left(\dfrac{1}{2}\theta_2\right)}{1 + \sqrt{\sin(\theta_1)}},
$$

or, at last,

$$
T_B = 2\sqrt{2}\sqrt{\frac{L}{g}} \left[\frac{\sin\left(\dfrac{1}{2}\theta_1\right)}{\sqrt{\sin(\theta_1)}} + \frac{\sin\left(\dfrac{90° - \theta_1}{2}\right)}{1 + \sqrt{\sin(\theta_1)}} \right].
$$

We can compare T_B to T_D by studying their ratio as a function of θ_1, i.e.,

$$
R = \frac{T_B}{T_D} = \sqrt{2} \left[\frac{\sin\left(\dfrac{1}{2}\theta_1\right)}{\sqrt{\sin(\theta_1)}} + \frac{\sin\left(45° - \dfrac{1}{2}\theta_1\right)}{1 + \sqrt{\sin(\theta_1)}} \right].
$$

A plot of R is given in figure 6.4, which shows that $R \leq 1$ for all θ_1 in the interval 0° to 90°, which means the bead always takes less time to descend along a broken path (even though it is the

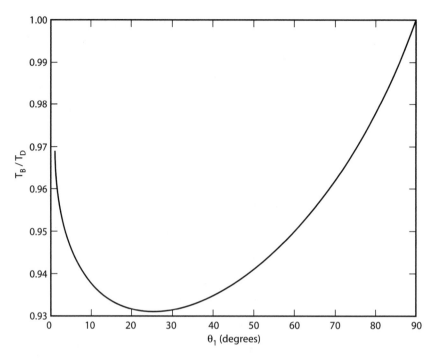

FIGURE 6.4. The two-segment broken line is almost as fast as the circle!

longer path) than along the direct path. (It is geometrically obvious that the broken path is longer than the direct path.) Only when $\theta_1 = 0°$ or $90°$ is $R = 1$, which is geometrically obvious since for both cases the broken path degenerates into the direct path. The plot shows that the time of descent is minimized when θ_1 is around 25°, although it is not a sharp minimum. A careful examination of the plot shows that the minimum value of R is 0.9313, i.e., at the minimum $T_B = 0.9313T_D$. Since $T_D = 1.0787T$, then at the minimum of figure 6.4, we have $T_B = (0.9313)(1.0787)T = 1.0046T$. With just a two-segment approximation to the circle, then, we can have a descent time less than $\frac{1}{2}$ of 1% greater than the circular descent time.

With this result in hand, $T_B < T_D$, Galileo then made his first mistake. He argued that the *double*-broken path (*DBEC*, shown in figure 6.5) would have a descent time even shorter than T_B. This

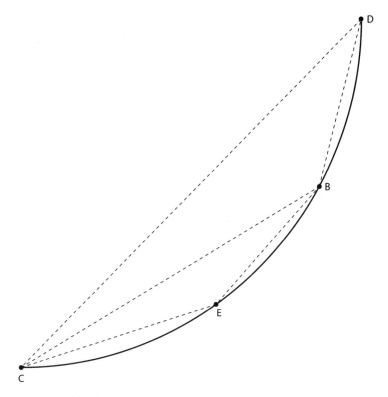

FIGURE 6.5. Galileo's mistake.

conclusion is correct, but his reasoning was not. He argued that, in terms of *time*, the single-broken path *DBC* is such that

$$DC > DB + BC,$$

and that the single-broken path *BEC* is such that

$$BC > BE + EC.$$

The first statement is of course true (we derived it!), but the second does not follow from our analysis because in the first analysis we assumed that the initial velocity is zero (as it *is* at *D*). But the initial velocity is *not* zero at *B*. By continuing to add more and more break points along the circular arc, Galileo concluded that the fastest way from *D* to *C* was along the circle itself, which is true (but, again, his reasoning was faulty). What some historians *think* he meant was that this is so *if* all the break points must be on the circle. Others think he meant the circle was the fastest descent curve of all possible curves from *D* to *C*. In fact, it is not, as the next section will demonstrate.

6.2 The Brachistochrone Problem

Once Galileo's original problem had focused attention on the general problem of gravitational descent, it was a natural question to then ask what *is* the curve of swiftest descent? Mathematicians of the caliber of the Bernoulli brothers, Newton, and Leibniz knew that Galileo's analysis had *not* established that it is a circular arc. What if, they asked, the broken-line approximation to the descent curve was no longer constrained to have all of its endpoints on a circular arc—perhaps then there could be an even "faster" curve.

It was this problem, of determining what is called the *brachistochrone*, that Johann Bernoulli posed "to the most acute mathematicians of the entire world" in June 1696. (The name comes from the Greek *brachistos* (shortest) and *chronos* (time) and is due to Bernoulli. Leibniz preferred *tachystoptote*, from *tachystos* (swiftest) and *piptein* (to fall), but deferred to Bernoulli.) Notice that this is not a problem of "ordinary" calculus, where what is asked for is the particular *value* of a variable that minimizes a function of that variable. Rather, we are now to find the *function* (i.e., a particular *entire curve*) that minimizes some other function (the so-called *functional*) whose independent "variable" takes on "values" from the set of all possible curves connecting two given points (in the brachistochrone problem, the "other function" is the descent time). This is an entirely new sort of minimization problem, and its solution initiated a new branch of mathematics—the calculus of variations.

Bernoulli's challenge to find the brachistochrone was accepted by some of the great mathematical minds of the day, but it was Bernoulli's own original solution that was the most beautiful and compelling, using a brilliant application of Fermat's principle of least time and Snell's law. In a 1697 letter, Bernoulli claimed to have had, however, no prior knowledge of Galileo's work on gravitational descent, and perhaps he was being honest. It strikes me as most unlikely, however, that Bernoulli could really have been so unaware—

it wasn't as if Galileo had published his circular descent analysis in some obscure journal. *Discourses* was a famous book! In addition, it is known that Johann Bernoulli had an extraordinarily jealous nature, and hated to share credit in mathematical work. We've already seen that side of him in the affair over who really wrote de L' Hospital's calculus book, and it was on display again in a later, very ugly business with his own son, Daniel, an accomplished mathematician in his own right. Daniel's important book *Hydrodynamica* was published in 1738, just as his father's similarly titled book *Hydraulica* was being published. Rather than being proud of his son, Johann claimed *he* had priority, even though he knew Daniel had actually finished his writing several years earlier. If Johann would deny his own son honest credit, then it is difficult to believe he would worry much about denying the long-dead Galileo any credit for motivating the brachistochrone problem.

Still, while Johann Bernoulli apparently had a serious problem with intellectual honesty, it cannot be denied he was a genius. His solution for the brachistochrone would alone insure his mathematical fame. Here's how he did it. From Snell's law, as correctly explained by Fermat's invoking of the principle of least time (see section 4.6), we have

$$\frac{\sin(\theta_1)}{v_1} = \frac{\sin(\theta_2)}{v_2} = \text{constant}$$

for a light ray traveling in the two mediums from B to A (speed v_1 and v_2 in the upper and lower mediums, respectively), shown in figure 6.6. That figure is similar to figure 4.10 (here I have written θ_1 and θ_2 for θ_i and θ_r, respectively), where it was understood that $v_2 < v_1$ (the upper medium, 1, is less dense than the lower medium, 2, as would be the case for medium 1 as air and medium 2 as water). We could, however, simply reverse the path of the ray to get figure 6.7, which is just figure 6.6 flipped over. Snell's law is still true for Figure 6.7, of course, just as written above.

Now, imagine that instead of just the two mediums of figure 6.7, there are a great many layered mediums, each *less* dense than the layer above it. Then the light ray's speed *increases* as it penetrates the layers in the downward direction, and the ray bends ever more away from the vertical, as does the ray path illustrated in figure 6.8.

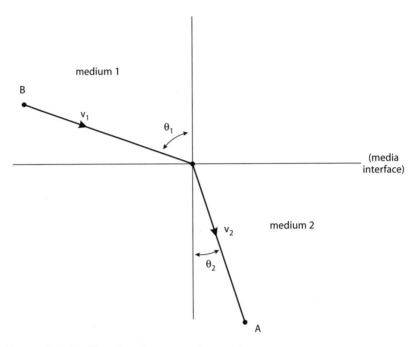

FIGURE 6.6. Snell's refraction geometry, again.

As we let the number of layers increase (and the thickness of each layer decrease) without bound, the path becomes a smooth curve, and at *every point* along this curve we will have

$$\frac{\sin(\theta)}{v} = \text{constant}.$$

Bernoulli's brilliant insight into how to solve the minimum-descent-time problem was to turn the above argument on its head. That is, if the above condition is the *result* of assuming minimum travel (descent) time (Fermat's principle of least time for light), then *starting* with the above condition should *result* in the curve of minimum descent time.

Therefore, as shown in figure 6.9, I have sketched the curve of minimum descent time (whatever it is!) from B (the origin) to A, with θ as the angle between the tangent at an arbitrary point (x, y) on the curve and the vertical. Notice that the positive y-axis points *downward* because we are studying a *falling* bead. At the arbitrary

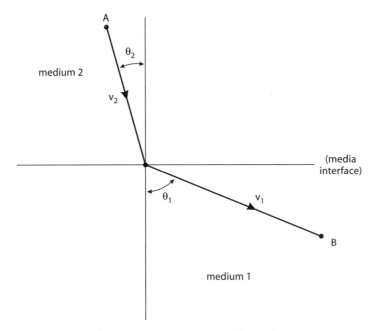

FIGURE 6.7. Snell's refraction geometry, again (flipped).

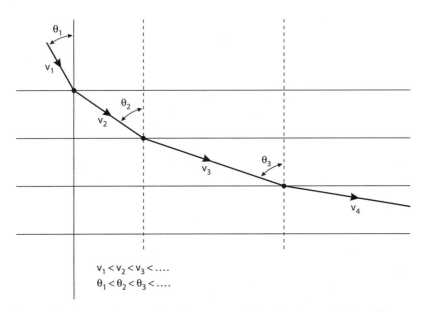

FIGURE 6.8. Layered approximation to a variable density optical medium.

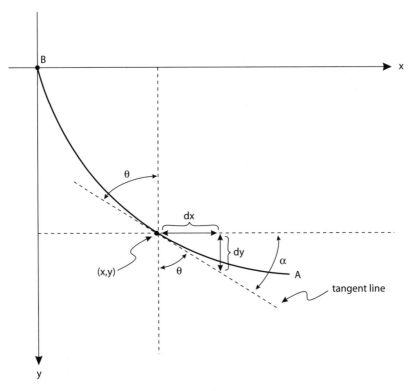

FIGURE 6.9. Geometry of Bernoulli's solution.

point (x, y) the speed of the descending bead along the curve is v. If we assume, as in Galileo's original analysis, that the bead starts its descent from B with zero initial speed, then conservation of energy says (for a bead with mass m) that the loss of potential energy (mgy) equals the gain in kinetic energy $\left(\frac{1}{2} mv^2\right)$, and so, after falling through a vertical distance of y, the speed of the bead is

$$v = \sqrt{2gy}.$$

So, Bernoulli's ingenious approach to the brachistochrone problem is "simply" to imagine that the "speed of light" in a variable-density optical medium is $\sqrt{2gy}$ and to find the path a ray of light will follow, because light takes the least-time path. This solution could only have occurred to a mind equally at home with mathematics *and* physics. Mathematical skill alone would not have been enough.

As de L'Hospital wrote to Bernoulli in a letter dated June 15, 1696, "This problem [of minimum descent time] seems to be one of the most curious and beautiful that has ever been proposed, and I would very much like to apply my efforts to it, but for this it would be necessary that you reduce it to pure mathematics, since physics bothers me."

From the geometry of figure 6.9 it is clear that

$$\sin(\theta) = \cos(\alpha) = \frac{1}{\sec(\alpha)} = \frac{1}{\sqrt{1 + \tan^2(\alpha)}} = \frac{1}{\sqrt{1 + \left(\dfrac{dy}{dx}\right)^2}}$$

$$= \frac{1}{\sqrt{1 + (y')^2}}.$$

Therefore,

$$\frac{\sin(\theta)}{v} = \text{constant} = \frac{\dfrac{1}{\sqrt{1 + (y')^2}}}{\sqrt{2gy}}.$$

Squaring the second equality gives

$$2gy\left[1 + (y')^2\right] = \text{constant},$$

or, finally, with C a constant, we arrive at the (nonlinear) differential equation for the curve of minimum descent time:

$$y\left[1 + \left(\frac{dy}{dx}\right)^2\right] = C.$$

Nonlinear differential equations are generally not easy to solve analytically (with each new one requiring, it seems, its own unique "trick"), but we can solve this one for y in the following way. Taking advantage of Leibniz's notational advantage over that of Newton's, and treating the differentials dx and dy as algebraic quantities, we can solve for dx to get

$$dx = dy\sqrt{\frac{y}{C - y}}.$$

Next, making the change of variable to φ (notice that $\varphi = 0$ when $y = 0$), where

$$\tan(\varphi) = \sqrt{\frac{y}{C - y}} = \frac{\sin(\varphi)}{\cos(\varphi)},$$

we have

$$\frac{y}{C - y} = \frac{\sin^2(\varphi)}{\cos^2(\varphi)},$$

$$y \cos^2(\varphi) = C \sin^2(\varphi) - y \sin^2(\varphi),$$

$$y \cos^2(\varphi) + y \sin^2(\varphi) = y = C \sin^2(\varphi).$$

Differentiation of the last equality with respect to φ gives

$$\frac{dy}{d\varphi} = 2C \sin(\varphi) \cos(\varphi),$$

and so $dy = 2C \sin(\varphi) \cos(\varphi) d\varphi$, which says

$$dx = 2C \sin(\varphi) \cos(\varphi) \sqrt{\frac{y}{C - y}} d\varphi = 2C \sin(\varphi) \cos(\varphi) \tan(\varphi) d\varphi.$$

Or, as $\cos(\varphi) \tan(\varphi) = \sin(\varphi)$, we have (using a trigonometric double-angle identity)

$$dx = 2C \sin^2(\varphi) d\varphi = C[1 - \cos(2\varphi)] \, d\varphi.$$

This last expression we can integrate by inspection: with C_1 as the constant of indefinite integration, we arrive at

$$x = C\left[\varphi - \frac{\sin(2\varphi)}{2}\right] + C_1 = \frac{1}{2} C[2\varphi - \sin(2\varphi)] + C_1.$$

We can determine the value of C_1 by inserting the coordinates of the point B at which the descent begins, that is, the origin $x = y = 0$. Or, equivalently, $x = \varphi = 0$. Then, C_1 is obviously zero and so

$$x = \frac{1}{2} C[2\varphi - \sin(2\varphi)].$$

Earlier we also found that $y = C \sin^2(\varphi) = C[1 - \cos^2(\varphi)]$ and so, again from a trigonometric double-angle identity, we have

$$y = \frac{1}{2}C[1 - \cos(2\varphi)].$$

As our final step, to make the equations as simple-appearing as possible, I'll replace the constant $\frac{1}{2}C$ with simply a, and make the change of variable $\beta = 2\varphi$. Then, at last, we arrive at the so-called *parametric equations* for the minimum-descent-time curve, or brachistochrone:

$$
\boxed{\begin{aligned}
x &= a[\beta - \sin(\beta)] \\
y &= a[1 - \cos(\beta)]
\end{aligned}}
\,.
$$

This result greatly surprised Bernoulli, who recognized these equations as describing a previously known (for at least a century) curve, the *cycloid* (a name coined by Galileo in 1599), which is the curve traced by a point (starting at the origin) on the circumference of a wheel, with radius a, rolling without slipping along the x-axis. Although it seems incredible that the cycloid could have been overlooked by the ancient mathematicians, it appears that the first time it was discussed in print was in 1501, in the work of the French mathematician Charles Bouvelles (1470–1553). You can find more discussion in the paper by E. A. Whitman, "Some Historical Notes on the Cycloid" (*American Mathematical Monthly*, May 1943, pp. 309–15).

The cycloid equations do not directly connect x and y, but rather link them together via the parameter β. We can thus simply vary β, calculate x and y for each of many different values of β, and arrive at the x, y plot of the cycloid. Figure 6.10 shows such a plot, for $a = 1$, in the interval $0 \le \beta \le 2\pi$. It should be clear that a is simply a scale factor and, as we make a smaller or larger, the curve shrinks or inflates, respectively. We can make the cycloid, starting at the origin, pass through any given point $(x > 0, y > 0)$ by simply picking the constant a properly (start with $a = 0$ and then increase it, i.e., "inflate" the cycloid, until it passes through the given point). This should make it obvious, too, that there is a *unique* value of a that does this. Thus, the brachistochrone joining two points is unique, and it is an *inverted* section of the arch of a cycloid.

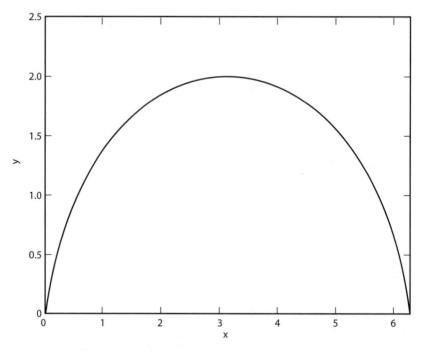

FIGURE 6.10. The cycloid $a = 1$.

You should not think of the parametric representation of a curve as being something less than desirable, as somehow being less useful than a direct expression of y in terms of x. We will not find ourselves at any disadvantage with a parametric representation. For example, if we want to know the slope of the cycloid at some point, we simply use the chain rule to calculate

$$\frac{dy}{dx} = \frac{dy}{d\beta} \cdot \frac{d\beta}{dx} = \frac{dy}{d\beta} \bigg/ \frac{dx}{d\beta} = \frac{\sin(\beta)}{1 - \cos(\beta)}.$$

There can be occasions, in fact, where the parametric representation is the *only* proper way to formulate a problem. For example, in appendix F you'll find the derivation of an expression for the area inside a closed, non-self-intersecting curve:

if the parametric equations of the curve C are $x = x(t)$ and $y = y(t)$, then

$$\text{area enclosed by } C = \frac{1}{2} \int_0^T \left(y \frac{dx}{dt} - x \frac{dy}{dt} \right) dt,$$

where C is imagined to be the *clockwise* path traversed by a moving point, starting at time $t = 0$ at some place and returning to that initial place at time $t = T$. This result will be crucial to the solution of the ancient isoperimetric problem discussed in chapter 2 (what figure of given perimeter encloses the maximum area?), and which we will finally be able to do in this chapter.

Johann Bernoulli's brother Jacob (1654–1705), Leibniz, and Newton also submitted solutions in response to Johann's challenge. Bernoulli's challenge to Newton, in particular, was not really a friendly one. Bernoulli had taken Leibniz's side in the dispute over who was the "true" discoverer of the calculus, and he meant to embarrass Newton by showing that he was unable to solve a problem that both Bernoulli and Leibniz had already solved. As Bernoulli stated in the public announcement of the brachistochrone problem, "so few have appeared to solve our extraordinary problem, even among those who boast that through special methods, which they commend so highly, they have not only penetrated the deepest secrets of geometry but also extended its boundaries in marvelous fashion; although their golden theorems which they imagine were known to no one, have been published by others long before."

Newton was not amused by this; as he later stated, "I do not love to be dunned and teased by foreigners about Mathematical things." Newton quickly set about answering Bernoulli's challenge and, according to second-hand accounts, solved the problem in a single night using a then unknown method (but see the box in section 6.4). Newton's "solution," however, is simply a description for how to construct the minimum-descent-time cycloid, with no explanation for how he arrived at that curve as the brachistochrone. The construction was published anonymously in the *Philosophical Transactions of the Royal Society* of January 1697 (backdated by his editor/friend Edmond Halley, as Newton actually *first* read aloud his "solution" at a meeting of the Royal Society on February 24, 1697).

A famous story about the anonymous publication is that, after reading it, Johann claimed he knew the unnamed author was Newton because he "recognized the lion by his paw." For once in his life Johann Bernoulli, despite his bias against Newton, was gracious to a competing mathematician working on the same problem, perhaps because in this case Bernoulli clearly had priority. [However, for a more sympathetic view of Johann Bernoulli's relationships with competing mathematicians see the old but still valuable paper by Constantin Carathéodory, "The Beginning of Research in the Calculus of Variations" (*Osiris*, 1938, pp. 224–40)].

The brachistochrone has a second remarkable property, in addition to being the curve of minimum descent time. In 1656 the Dutch mathematical physicist Christiaan Huygens (1629–95) constructed the first successful pendulum clock, which he knew had a period slightly dependent on the amplitude of the pendulum swing. To achieve complete independence, i.e., to invent the so-called *isochronous* pendulum clock, Huygens inserted curved metal surfaces at the suspension point on each side of the flexible cord that (along with a weight at the end) served as the pendulum. These surfaces forced the pendulum cord to deviate from being straight as it swung back and forth, in just such a way as to make the period independent of the amplitude of the swing. In his 1673 masterpiece, *Horologium Oscillatorium* (*The Pendulum Clock*), Huygens showed that the curved constraint surfaces should be cycloidal arcs (he had actually known this since the end of 1659). That would force the swinging weight to follow a cycloidal path (a mathematician would say that Huygens had discovered that the *involute* of a cycloid is another cycloid), which was known to be isochronous, i.e., a bead undergoing gravitational descent along a cycloidal curve takes the *same time* to reach the bottom of the curve, no matter where it starts its descent (this is shown in the next section). This means the brachistochrone is also a *tautochrone* (from the Greek *tauto*, the same, and of course, *chronous*, time), a discovery that so pleased Huygens he said it was "the most fortunate finding which ever befell me." In actual practice, however, the friction between the curved metal surfaces and the pendulum cord resulted in a bigger source of timekeeping error than was the original amplitude-period dependency.

As I mentioned earlier, Bernoulli was astonished to learn the brachistochrone is a cycloid, and so, when he revealed his derivation in January 1697, he first discussed Huygen's cycloid and its tautochronous property and then stated, "you will be petrified with astonishment when I say that precisely this same cycloid . . . is our required brachistochrone. . . . Nature always tends to act in the simplest way [certainly Bernoulli would say this, since he had used Fermat's principle of least time in arriving at his solution], and so it here lets one curve serve two different functions."

6.3   Comparing Galileo and Bernoulli

Now that we have the analytic form of the true minimum-descent-time curve, the next natural question to ask is how much faster is it than Galileo's *circular* descent curve? We found in section 6.1 that, on a quarter circle of radius L, it takes the time T for the bead to make the descent, where

$$T = 1.8541\sqrt{\frac{L}{g}}.$$

Galileo didn't actually calculate this result, but he came close to it, and so I'll now write T as T_G. What we want to calculate now is T_B, the time to fall along the brachistochrone curve from $(0, 0)$ to (L, L). (Note carefully: this T_B is *not* the T_B of section 6.1!) Everything we've done so far tells us $T_B < T_G$. Let's see by how much.

If we define s as the distance from the origin to the arbitrary point (x, y) on the descent curve, as measured along the curve, then, as argued before from the conservation of energy, we have

$$v = \frac{ds}{dt} = \sqrt{2gy},$$

where, from the Pythagorean theorem, we have the differential arc length ds along the curve as $ds = \sqrt{(dx)^2 + (dy)^2}$. Thus,

$$v = \frac{\sqrt{(dx)^2 + (dy)^2}}{dt} = \frac{dx\sqrt{1 + \left(\dfrac{dy}{dx}\right)^2}}{dt},$$

and so

$$dt = \frac{dx\sqrt{1 + \left(\frac{dy}{dx}\right)^2}}{v} = \frac{dx\sqrt{1 + \left(\frac{dy}{dx}\right)^2}}{\sqrt{2gy}}.$$

Integrating, where, as t goes from 0 to T_B we have x go from 0 to L,

$$\int_0^{T_B} dt = \int_0^L \sqrt{\frac{1 + \left(\frac{dy}{dx}\right)^2}{2gy}} \, dx = T_B.$$

Because we already have the equations relating y and x (the parametric equations of the cycloid), we can now directly evaluate this integral, as I'll do next. But first, notice that we have arrived at this integral (the so-called *functional*) without using our knowledge of the specific relationship between y and x. Indeed, the general approach of the calculus of variations (which we'll take up in the next section) does not require that knowledge, but instead derives the brachistochrone by determining the *function $y(x)$* that minimizes the time functional. For now, however, let's evaluate T_B directly.

From the boxed parametric equations for the brachistochrone given in the previous section, we have

$$\frac{dx}{d\beta} = a[1 - \cos(\beta)]$$

$$\frac{dy}{d\beta} = a \, \sin(\beta).$$

We have $\beta = 0$ when $x = 0$ from the definition of β, and let's further suppose that $\beta = \hat{\beta}$ when $x = L$. Then,

$$T_B = \int_0^L \sqrt{\frac{1 + \left(\frac{dy}{dx}\right)^2}{2gy}} \, dx = \int_0^L \sqrt{\frac{(dx)^2 + (dy)^2}{2gy}}$$

$$= \int_0^{\hat{\beta}} \sqrt{\frac{a^2[1 - \cos(\beta)]^2 + a^2 \sin^2(\beta)}{2ga \, [1 - \cos(\beta)]}} \, d\beta = \int_0^{\hat{\beta}} \sqrt{\frac{2a^2[1 - \cos(\beta)]}{2ga[1 - \cos(\beta)]}} \, d\beta$$

$$= \hat{\beta}\sqrt{\frac{a}{g}}.$$

To find $\hat{\beta}$ (which certainly must be greater than zero) and a, we use the fact that the brachistochrone ends at (L,L)—remember, as in figure 6.9, we are thinking of the positive y-axis as increasing *downward*. Thus,

$$L = a\big[\hat{\beta} - \sin(\hat{\beta})\big]$$

$$L = a\big[1 - \cos(\hat{\beta})\big],$$

and so

$$a = \frac{L}{\hat{\beta} - \sin(\hat{\beta})} = \frac{L}{1 - \cos(\hat{\beta})}.$$

The second equality is equivalent to solving the equation

$$f(\beta) = \beta + \cos(\beta) - \sin(\beta) - 1 = 0.$$

A plot of $f(\beta)$ is shown in figure 6.11, which tells us there is just one positive solution to $f(\beta) = 0$. Using the Newton-Raphson iterative method discussed in section 4.5, it is easy to calculate that solution to be $\hat{\beta} = 2.412$ radians. Thus,

$$a = \frac{L}{1 - \cos(2.412)} = \frac{L}{2.412 - \sin(2.412)} = 0.5729\,L,$$

and so

$$T_B = 2.412\sqrt{\frac{0.5729\,L}{g}} = 1.8257\sqrt{\frac{L}{g}},$$

which is, indeed, less than T_G. But only by about 1.5%. Galileo's quarter-circle *is* pretty close to being the brachistochrone.

We can show the isochronous property of the cycloid as follows. For a cycloid starting at $(0,0)$ and ending at the *very bottom* of the cycloidal path, we have (from before)

$$T = \hat{\beta}\sqrt{\frac{a}{g}},$$

where now $\hat{\beta}$ is the value of β at the bottom. (For the brachistochrone joining $(0,0)$ to (L, L), the problem we just analyzed, the

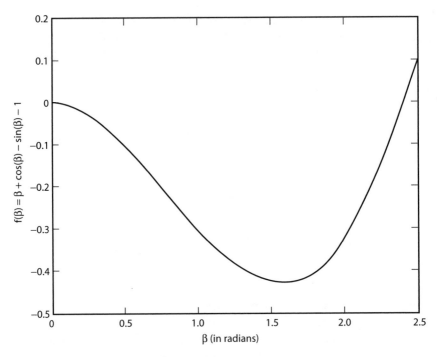

FIGURE 6.11. Estimating β when $x = L$.

point (L, L) is *not* the bottom of the cycloidal path). From the parametric equations of the cycloid, we see that this means $\hat{\beta} = \pi$: at the bottom, $x = \pi a$ and $y = 2a$ (take a look again at figure 6.10). Thus, the time required for a bead to slide from top to bottom is

$$T = \pi \sqrt{\frac{a}{g}}.$$

If the fall along the cycloid does not start at $(0,0)$, however, but rather at some lower point (x_0, y_0) on the cycloid, then the speed of the descending bead, at the general point (x, y), is

$$v = \sqrt{2g(y - y_0)},$$

and so the time to reach the bottom is now given by

$$T' = \int_{x_0}^{\pi a} \sqrt{\frac{1 + \left(\dfrac{dy}{dx}\right)^2}{2g(y - y_0)}} \, dx.$$

The isochronous property discovered by Huygens says $T' = T$. Here's why.

Inserting dx and dy in terms of β, as we did before, and changing the integration limits to the appropriate values for β (let $\beta = \beta_0$ at (x_0, y_0)), we have

$$T' = \int_{x_0}^{\pi a} \sqrt{\frac{(dx)^2 + (dy)^2}{2g(y - y_0)}}$$

$$= \int_{\beta_0}^{\pi} \sqrt{\frac{a^2[1 - \cos(\beta)]^2 + a^2 \sin^2(\beta)}{2g[\{a - a\cos(\beta)\} - \{a - a\cos(\beta_0)\}]}} \, d\beta$$

$$= \int_{\beta_0}^{\pi} \sqrt{\frac{2a^2[1 - \cos(\beta)]}{2ag\cos(\beta_0) - 2ag\cos(\beta)}} \, d\beta$$

$$= \sqrt{\frac{a}{g}} \int_{\beta_0}^{\pi} \sqrt{\frac{1 - \cos(\beta)}{\cos(\beta_0) - \cos(\beta)}} \, d\beta.$$

From the half-angle trigonometric identity

$$\sin\left(\frac{1}{2}\beta\right) = \sqrt{\frac{1 - \cos(\beta)}{2}},$$

we then have

$$T' = \sqrt{\frac{a}{g}} \int_{\beta_0}^{\pi} \frac{\sqrt{2}\sin\left(\frac{1}{2}\beta\right)}{\sqrt{\cos(\beta_0) - \cos(\beta)}} \, d\beta.$$

And from the half-angle identity

$$\cos\left(\frac{1}{2}\beta\right) = \sqrt{\frac{1 + \cos(\beta)}{2}},$$

we have $\cos(\beta) = 2\cos^2\left(\frac{1}{2}\beta\right) - 1$, and so

$$T' = \sqrt{\frac{a}{g}} \int_{\beta_0}^{\pi} \frac{\sqrt{2}\sin\left(\frac{1}{2}\beta\right)}{\sqrt{2\cos^2\left(\frac{1}{2}\beta_0\right) - 1 - 2\cos^2\left(\frac{1}{2}\beta\right) + 1}} \, d\beta$$

$$= \sqrt{\frac{a}{g}} \int_{\beta_0}^{\pi} \frac{\sin\left(\frac{1}{2}\beta\right)}{\sqrt{\cos^2\left(\frac{1}{2}\beta_0\right) - \cos^2\left(\frac{1}{2}\beta\right)}} \, d\beta.$$

If we now change the integration variable to

$$u = \frac{\cos\left(\frac{1}{2}\beta\right)}{\cos\left(\frac{1}{2}\beta_0\right)},$$

then

$$\frac{du}{d\beta} = -\frac{\sin\left(\frac{1}{2}\beta\right)}{2\cos\left(\frac{1}{2}\beta_0\right)},$$

and so the T' integral becomes

$$T' = \sqrt{\frac{a}{g}} \int_{1}^{0} \frac{-2\cos\left(\frac{1}{2}\beta_0\right)}{\sqrt{\cos^2\left(\frac{1}{2}\beta_0\right) - u^2 \cos^2\left(\frac{1}{2}\beta_0\right)}} \, du$$

$$= 2\sqrt{\frac{a}{g}} \int_{0}^{1} \frac{du}{\sqrt{1-u^2}}.$$

From integral tables, we find this integral is $\sin^{-1}(u)$, and so

$$T' = 2\sqrt{\frac{a}{g}} \left\{ \sin^{-1}(u) \Big|_{0}^{1} \right\} = 2\sqrt{\frac{a}{g}} \left\{ \sin^{-1}(1) - \sin^{-1}(0) \right\}$$

$$= 2\sqrt{\frac{a}{g}} \left\{ \frac{\pi}{2} - 0 \right\} = \pi\sqrt{\frac{a}{g}} = T,$$

as claimed.

With a knowledge of the chain rule in differentiation, we can actually derive the above integral easily, with no need for tables. In figure 6.12, I've drawn a right triangle such that the angle φ is given by $\sin(\varphi) = u$, i.e., $\varphi = \sin^{-1}(u)$. Thus,

$$\frac{d}{du}\sin^{-1}(u) = \frac{d\varphi}{du}.$$

From the chain rule, and the figure,

$$\frac{d}{du}\sin(\varphi) = \cos(\varphi)\frac{d\varphi}{du} = \sqrt{1-u^2}\frac{d\varphi}{du}.$$

But of course we also have

$$\frac{d}{du}\sin(\varphi) = \frac{du}{du} = 1.$$

Thus,

$$1 = \sqrt{1-u^2}\frac{d\varphi}{du}, \quad \text{or} \quad \frac{d\varphi}{du} = \frac{1}{\sqrt{1-u^2}},$$

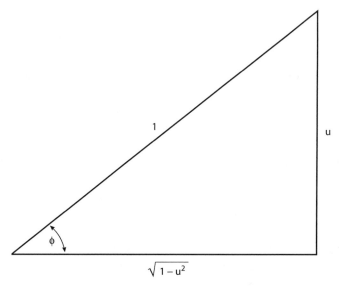

FIGURE 6.12. Differentiating the inverse sine function.

and so

$$\frac{d}{du}\sin^{-1}(u) = \frac{1}{\sqrt{1-u^2}}.$$

Integrating both sides then immediately gives us

$$\int \frac{du}{\sqrt{1-u^2}} = \sin^{-1}(u).$$

Bernoulli was certainly correct in saying the cycloid has a fascination all of its own, even for nonmathematicians. Its isochronous property, for example, received attention in, of all places, a famous work of fiction, Herman Melville's 1851 classic whaling story *Moby-Dick*. In chapter 96 ("The Try-Works"), where the book's narrator (remember him?—"Call me Ishmael") is describing how the try-pots of the ship *Pequod* are cleaned (a try-pot is an enormous iron cauldron used to reduce whale blubber to liquid oil), we read the following passage:

> ... an American whaler is outwardly distinguished by her try-works. . . . The try-works are planted between the foremast and main-mast, the most roomy part of the deck. The timbers beneath are of a peculiar strength, fitted to sustain the weight of an almost solid mass of brick and mortar, some ten feet by eight square, and five in height. The foundation does not penetrate the deck, but the masonry is firmly secured to the surface by ponderous knees of iron bracing it on all sides, and screwing it down to the timbers. On the flanks it is cased with wood, and at top completely covered by a large, sloping, battened hatchway. Removing this hatch we expose the great try-pots, two in number, and each of several barrels' capacity. When not in use, they are kept remarkably clean. Sometimes they are polished with soapstone and sand, till they shine within like silver punch bowls. During the night watches some cynical old sailors will crawl into them and coil themselves away there for a nap. While employed in polishing them—one man in each pot, side by side—many confidential communications are carried on, over the iron lips. *It is a place also for profound*

mathematical meditation. It was in the left hand try-pot of the Pequod, with the soapstone diligently circling around me, that I was first indirectly struck by the remarkable fact, that in geometry all bodies gliding along the cycloid, my soapstone for example, will descend from any point in precisely the same time [my emphasis].

The problem of determining the curve of swiftest descent is not simply one of historical interest. It has reappeared over the centuries in various forms, right up to modern times (it represents the ultimate in fast roller coaster rides, for example, especially right at the start with a *vertical* drop!), and it continues to capture the imagination. For example, in 1966, Paul W. Cooper, an industrial mathematician, published a short note in the *American Journal of Physics* ("Through the Earth in Forty Minutes," January, pp. 68–70). In his paper, Cooper pointed out that the gravitational field of the *interior* of the Earth would allow "falling through" a frictionless, straight tunnel connecting *any* two points on the surface of the planet in the same time interval of 42.2 minutes. Cooper imagined "a transportation system without timetables wherein the world's cities are linked with chords and where the departure time is universally on the hour and arrival time forty-two minutes later. Such a chord link between Boston and Washington, D.C., would involve a maximum penetration of about 50 miles below the Earth's surface." [For a somewhat more realistic tunnel transportation system, see the earlier paper by L. K. Edwards, "High-Speed Tube Transportation" (*Scientific American*, August 1965, pp. 30–40).]

Straight tunnels do not define the fastest travel times, however, and Cooper also wrote "One might want, in fact, the actual minimal-time path. . . . This is more complex than the classic brachistochrone problem in that here [by *here* Cooper means *inside* the Earth] the gravitational field is radial instead of rectangular, and it is not uniform." The brachistochrone tunnel connecting Los Angeles and New York City, for example, has a travel time of just 28 minutes—but it comes with a high price. The curved tunnel dips 1,000 miles below the surface! See J. E. Prussing, "Brachistochrone-tautochrone Problem in a Homogeneous Sphere" (*American Journal of Physics*, March 1976, pp. 304–5).

Cooper's paper on straight tunnels was noticed by *Time* magazine (February 11, 1966, pp. 42–43) and given a science fiction flavor for its popular audience, which prompted a number of physicists to write the *AJP* to say the whole idea was actually an old idea (see the replies in the August 1966 issue of the *American Journal of Physics*, pp. 701–4). Indeed, one writer traced it back to a paper delivered to the French Association for the Advancement of Science, in 1883! That paper may well have been the inspiration for Lewis Carroll, who used the concept in chapter 7 of his novel *Sylvie and Bruno Concluded* ten years later.

Three years after Cooper's paper appeared, a brief solution to the brachistochrone problem inside the Earth was given in the *American Mathematical Monthly* ("Fast Tunnels through the Earth," June–July 1969, pp. 708–9). That solution uses the modern calculus of variations approach. Twelve years after that, P. K. Aravind, a chemist (!) at the University of California/Santa Barbara, showed how to use Bernoulli's original optical analogy approach to solve the interior problem ("Simplified Approach to Brachistochrone Problems," *American Journal of Physics*, September 1981, pp. 884–86). And finally, the brachistochrone problem can be solved in closed form even if the additional complication of friction is included (we've ignored that important reality in all that we've done in this chapter). It is not a trivial exercise, however, and I'll simply refer you to the paper by N. Ashby et al., "Brachistochrone with Coulomb Friction," *American Journal of Physics*, October 1975, pp. 902–6.

In this section we saw, for the first time, the formula for the length of a curve traced out by a moving point. If that motion is described by $x = x(t)$ and $y = y(t)$, then the path length traveled over the time interval $t = 0$ to $t = T$ is

$$\int_0^T \sqrt{\left(\frac{dx}{dt}\right)^2 + \left(\frac{dy}{dt}\right)^2}\, dt.$$

An interesting extrema problem uses this result: if Tiger Woods wants to hit a golf ball for maximum range, then we showed in section 5.4 that he should drive the ball off of its tee at a 45° angle. But suppose instead that he wants to drive the ball for maximum distance through space, i.e., for maximum *trajectory length?* Then the angle is not 45°, but rather the not-so-obvious 56.466°. You can find it all worked out in the paper by Ze-Li Dou and Susan G. Staples, "Maximizing the Arclength in the Cannonball Problem" (*The College Mathematics Journal*, January 1999, pp. 44–45).

6.4   The Euler-Lagrange Equation

The brachistochrone problem is generally accepted by historians as marking the beginning of the calculus of variations. This, despite the fact that Bernoulli's solution, using Fermat's principle of least time and Snell's law, does *not* use the methods of that yet to be developed subject. The reason for this is because it was quickly understood by all that, while undeniably brilliant, Bernoulli's solution by optical analogy was too specialized, with no hope of being extended to other such questions, e.g., to the ancient isoperimetric question, discussed in chapter 2, of what closed curve, of given length, encloses the maximum area? What was needed was a general theory to attack such problems, and the brachistochrone *problem* itself, not Bernoulli's particular solution of it, was the spark that initiated the search for that theory.

Sometimes one does read of an earlier problem that is said to actually be the *first* such problem in the calculus of variations, but its history is a murky one indeed. This is the question, briefly mentioned in Newton's *Principia* (in 1687, nine years before Bernoulli's brachistochrone challenge), of what solid of revolution would experience the least resistance to motion through a medium with certain physical properties (e.g., a ship's hull in water)? As did his later brachistochrone "solution," Newton's answer to the minimum-resistance problem appeared in the original Latin printing of the *Principia* as just that; an *answer* with no derivation (as the Scholium to Proposition 34 of Book 2). This has led some modern writers to conclude (oddly and without justification, in my opinion) that Newton had no proofs! See, for example, the paper by Robert Weinstock, "Isaac Newton: Credit Where Credit Won't Do," and the replies to it, in *The College Mathematics Journal* (May 1994, pp. 179–222). Neither Professor Weinstock nor his critics seem to be aware of the fact that when the English translation of the *Principia* appeared in 1729, the minimum-resistance solid *was* treated analytically (those calculus-based arguments were included without attribution, but it was established in 1888 that they were, indeed, from Newton—see the following box).

The analytical treatment appearing in the 1729 edition of the *Principia* of the minimum resistance solid, discovered in Newton's papers in 1888, was prompted by a request he received from a reader of the original Latin printing of the *Principia*. Newton replied with the requested analysis in a letter dated July 14, 1694, to David Gregory (1659–1708), two years *before* Bernoulli's challenge; the method used is easily extended to the brachistochrone problem. Gregory, a Scot who ended his career as the Savilian Professor of Astronomy at Oxford, had a reputation as a not very outstanding mathematician. His entry in the *Dictionary of Scientific Biography* tells us, for example, that "the impression gained from his printed work [is] that a modicum of talent, effectively lacking originality, was stretched a long way." The last paragraph of that same entry also makes it clear, however, that Gregory has all math historians in his debt: "In retrospect, Gregory's true role in the development of seventeenth-century science was not that of original innovator but that of custodian of certain precious papers and verbal communications passed to him . . . as privileged information, by Newton."

It seems clear (to me, at least) that Newton simply thought both the minimum-resistance-solid problem and the minimum-descent-time curve problem to be interesting but not worthy of lengthy elaboration. This remarkable pair of decisions simultaneously illustrates both his monumental genius as well as an even more monumental mistake in judgment! Without Gregory's request, we might well never have learned the details of Newton's solutions. You can find discussions of Newton's "missing" solutions in H. W. Turnbull, *The Mathematical Discoveries of Newton* (Blackie & Son 1945, pp. 39–42), in Herman H. Goldstine, *A History of the Calculus of Variations from the*

17th through the 19th Century (Springer-Verlag 1980, pp. 7–29), and in I. Bernard Cohen, "Isaac Newton, the Calculus of Variations, and the Design of Ships," in *For Dirk Struik* (D. Reidel 1974, pp. 169–87).

As with Bernoulli's solution to the brachistochrone problem, Newton's solution to the minimum-resistance solid (using a method easily extended to the brachistochrone problem, which means he surely *did derive* the cycloid solution) is *not* the proper basis for attacking other functional problems in general. The development of a general theory began with the Swiss genius Leonhard Euler (1707–83), a student of Bernoulli who went on to exceed his mentor.

What I'll do next, then, is derive the basis for just such a general approach to these problems, the so-called Euler-Lagrange equation. The presentation that follows is the modern one found in textbooks today, and is fairly close to the way it was first done by the French-Italian mathematical-physicist Joseph Louis Lagrange (1736–1813). The equation was known to Euler by 1736, but today it is universally derived in the same way that Lagrange did it (in a 1755 letter to Euler when Lagrange was, yes, just nineteen!), which Euler enthusiastically adopted as the superior approach. Lagrange's use of a "variational" technique prompted Euler to coin the name *calculus of variations*. For the historically minded, excellent papers to read for the detailed history are G. A. Bliss, "The Evolution of Problems of the Calculus of Variations" (*American Mathematical Monthly*, December 1936, pp. 598–609) and Craig G. Fraser, "Isoperimetric Problems in the Variational Calculus of Euler and Lagrange" (*Historia Mathematica*, February 1992, pp. 4–23).

The simplest form of our general, fundamental problem is easy to state: find the function $y(x)$ that minimizes the integral (or *functional*)

$$J = \int_{x_1}^{x_2} F\left\{x, y(x), y'(x)\right\} \, dx,$$

where x_1 and x_2 are given, the function F is given, and $y'(x) = d/dx\, y(x)$. Many of the classical problems of the calculus of variations can be put in this form. For example, recall from the previous section the expression for T_B, the descent time along the curve $y = y(x)$ from $(0,0)$ to (L,L):

$$T_B = \int_0^L \sqrt{\frac{1 + (y')^2}{2gy}}\, dx.$$

We were able, there, to directly evaluate the *minimum* descent time (the minimum of the integral) because we already knew from other considerations what the equations for the minimum-descent-time curve are (the brachistochrone is a cycloid). Soon, however, we will redo this problem by a direct minimization of the integral. In this particular case, we have

$$F\,\{x, y, y'\} = \frac{1}{\sqrt{2g}} \cdot \sqrt{\frac{1 + (y')^2}{y}}.$$

In this case, as will be true all through this chapter, F is a known function of x, y, and y', but y is *not* a known function of x. Indeed, that is our problem: what *is* $y = y(x)$ to minimize J? In this book we are mostly interested in the pioneering problems of extrema, and the above integral J is *almost* all we need to consider. I say *almost* because I will eventually make two extensions to the above problem statement, but I'll save them for later. So, let's begin.

In figure 6.13, I've drawn the curve defined by $y = y(x)$, in the interval $x_1 < x < x_2$, where we will take that $y(x)$ to be the actual solution curve we are after. Around it, as the dashed curve, is

$$Y(x) = y(x) + \varepsilon\mu(x),$$

where ε is any constant and $\mu(x)$ is an *almost* arbitrary (but always differentiable, as we'll also assume $y(x)$ to be) function. I say *almost* because we will put two constraints on $\mu(x)$; it must vanish at the endpoints, i.e., $\mu(x_1) = \mu(x_2) = 0$. You'll see why, soon, this is a desirable property for $\mu(x)$. Notice, too, that $Y(x) = y(x)$ if $\varepsilon = 0$, which will be useful to remember in just a bit. That is, $Y(x)$ is a perturbed version of the solution $y(x)$, and $\varepsilon\mu(x)$ is the *variation* of $Y(x)$ around $y(x)$.

Because J will, in general, depend on the value of ε, we can write our formulation of the general problem as: find the $y(x)$ that minimizes the integral

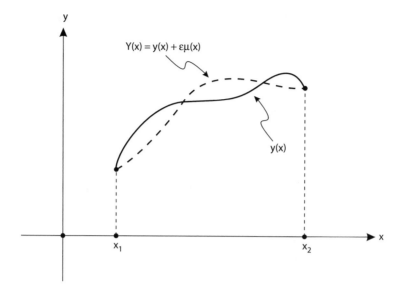

FIGURE 6.13. A true solution and a *variation* around it.

$$J(\varepsilon) = \int_{x_1}^{x_2} F\{x, Y(x), Y'(x)\} \, dx,$$

where

$$Y(x) = y(x) + \varepsilon\mu(x)$$
$$Y'(x) = y'(x) + \varepsilon\mu'(x).$$

$J = J(\varepsilon)$, since Y and Y' depend on ε. Now, since we have intention-ally constructed this formulation so that *by definition* $Y(x)$ collapses to the *solution* $y(x)$ when $\varepsilon = 0$, then the $J(\varepsilon)$ integral is minimized (*by definition!*) when $\varepsilon = 0$. Thus, it must be true that

$$\left.\frac{dJ}{d\varepsilon}\right|_{\varepsilon=0} = 0,$$

because this is the necessary (although of course *not* sufficient) con-dition for an extrema (e.g., a minimum) to exist. The distinction between an extrema being a maximum or a minimum will gener-ally be obvious from the physics of the particular problem we will be studying.

To proceed to our next step, I now need to use a result in calculus called *Leibniz's rule* for differentiating an integral. In the simple case we have, where the integration limits are not functions of ε, that rule reduces to the intuitively appealing result that the derivative of the integral is the integral of the derivative [the general rule, which *is* a bit more complicated, is nicely discussed by Marc Frantz, "Visualizing Leibniz's Rule" (*Mathematics Magazine*, April 2001, pp. 143–45)], and so

$$\frac{dJ}{d\varepsilon} = \frac{d}{d\varepsilon} \int_{x_1}^{x_2} F\{x, Y(x), Y'(x)\} \, dx = \int_{x_1}^{x_2} \frac{\partial F}{\partial \varepsilon} \, dx,$$

where $\partial F / \partial \varepsilon$ denotes the *partial* derivative of F. The *partial* refers to the fact that F is a function of variables other than just ε, i.e., x, Y, and Y'.

Using the chain rule, we can write $\partial F / \partial \varepsilon$ in terms of those other variables as

$$\frac{\partial F}{\partial \varepsilon} = \frac{\partial F}{\partial Y} \cdot \frac{\partial Y}{\partial \varepsilon} + \frac{\partial F}{\partial Y'} \cdot \frac{\partial Y'}{\partial \varepsilon} + \frac{\partial F}{\partial x} \cdot \frac{\partial x}{\partial \varepsilon}.$$

Since we have

$$\frac{\partial Y}{\partial \varepsilon} = \mu(x), \qquad \frac{\partial Y'}{\partial \varepsilon} = \mu'(x), \qquad \text{and} \quad \frac{\partial x}{\partial \varepsilon} = 0,$$

then

$$\frac{dJ}{d\varepsilon} = \int_{x_1}^{x_2} \left\{ \frac{\partial F}{\partial Y} \mu(x) + \frac{\partial F}{\partial Y'} \mu'(x) \right\} \, dx.$$

Since setting $\varepsilon = 0$ is equivalent to setting $Y = y$ and $Y' = y'$, we therefore have

$$\frac{dJ}{d\varepsilon}\bigg|_{\varepsilon=0} = 0 = \int_{x_1}^{x_2} \left\{ \frac{\partial F}{\partial y} \mu(x) + \frac{\partial F}{\partial y'} \mu'(x) \right\} \, dx.$$

To continue we next need to recall yet another result from calculus, the formula for integrating-by-parts that was developed in section 5.1: if $g(x)$ and $h(x)$ are two functions of x (in chapter 5, I used $u(x)$ in place of $g(x)$ and $f(x)$ in place of $h(x)$), then

$$\int_{x_1}^{x_2} g \, dh = \left(gh \Big|_{x_1}^{x_2} - \int_{x_1}^{x_2} h \, dg \right).$$

We can use this to integrate the second term of our expression for $dJ/d\varepsilon$. To do this, first set

$$g(x) = \frac{\partial F}{\partial y'}, \qquad dh = \mu'(x) \, dx.$$

Then,

$$\frac{dg}{dx} = \frac{d}{dx}\left(\frac{\partial F}{\partial y'} \right) \qquad \text{or } dg = \frac{d}{dx}\left(\frac{\partial F}{\partial y'} \right) dx$$

and

$$h(x) = \mu(x).$$

Thus,

$$\int_{x_1}^{x_2} \frac{\partial F}{\partial y'} \mu'(x) \, dx = \left(\frac{\partial F}{\partial y'} \mu(x) \Big|_{x_1}^{x_2} - \int_{x_1}^{x_2} \mu(x) \frac{d}{dx}\left(\frac{\partial F}{\partial y'} \right) dx \right).$$

However, since $\mu(x_1) = \mu(x_2) = 0$, we have

$$\left(\frac{\partial F}{\partial y'} \mu(x) \Big|_{x_1}^{x_2} = 0, \right.$$

and now you can see how convenient was that earlier stipulation that $\mu(x)$ vanish at both $x = x_1$ and $x = x_2$! This, then, lets us write

$$\frac{dJ}{d\varepsilon} \Big|_{\varepsilon=0} = 0 = \int_{x_1}^{x_2} \left\{ \frac{\partial F}{\partial y} \mu(x) - \mu(x) \frac{d}{dx}\left(\frac{\partial F}{\partial y'} \right) \right\} dx$$

$$= \int_{x_1}^{x_2} \mu(x) \left\{ \frac{\partial F}{\partial y} - \frac{d}{dx}\left(\frac{\partial F}{\partial y'} \right) \right\} dx.$$

We are now at the final step in deriving the Euler-Lagrange equation, a step so "obvious" to Lagrange that he zipped right through it. Later mathematicians thought this final step to be not quite so obvious, and so provided proofs for it, but I'll follow Lagrange and simply state the following so-called *fundamental lemma* of the calculus of variations (which I think *is* plausible):

if, for *arbitrary* $\mu(x)$, $\int_{x_1}^{x_2} \mu(x)H(x)dx = 0$,

then $H(x) = 0$, $x_1 < x < x_2$.

For us, this means

$$\frac{\partial F}{\partial y} - \frac{d}{dx}\left(\frac{\partial F}{\partial y'}\right) = 0$$

which I've put in a box because it *is* the famous Euler-Lagrange differential equation. Now, what do we *do* with it? That's what the rest of this chapter is about.

6.5   The Straight Line and the Brachistochrone

For our first application of the Euler-Lagrange equation, let's prove that the curve of minimum length connecting two given points in a plane is a straight line. Recall that the differential length ds along the curve $y = y(x)$ is

$$ds = \sqrt{(dx)^2 + (dy)^2} = \sqrt{1 + \left(\frac{dy}{dx}\right)^2}\, dx = \sqrt{1 + (y')^2}\, dx.$$

The total length of the curve connecting the points (x_1, y_1) and (x_2, y_2) is, therefore,

$$J = \int_{x_1}^{x_2} ds = \int_{x_1}^{x_2} \sqrt{1 + (y')^2}\, dx,$$

which means we have

$$F = \sqrt{1 + (y')^2} = \{1 + (y')^2\}^{1/2}.$$

Since F has no explicit dependence on y, we have

$$\frac{\partial F}{\partial y} = 0,$$

and we also see that

$$\frac{\partial F}{\partial y'} = \frac{1}{2}\{1 + (y')^2\}^{-1/2}\, 2y' = \frac{y'}{\sqrt{1 + (y')^2}}.$$

Inserting these two results into the Euler-Lagrange equation gives

$$\frac{d}{dx}\left[\frac{y'}{\sqrt{1 + (y')^2}}\right] = 0,$$

which immediately tells us that

$$\frac{y'}{\sqrt{1 + (y')^2}} = \text{constant.}$$

And *this*, in turn, immediately says that $y'(x) = $ constant, which means $y(x) = mx + b$, where m and b are constants. This is, of course, the equation for a straight line, with m and b selected to make the line pass through the given points (x_1, y_1) and (x_2, y_2). So, *at last*, we have mathematical proof of what we all knew all along! This is nonetheless an important result, helping to build our confidence in the Euler-Lagrange equation.

As a further confidence builder, let's solve another problem to which we also already know the answer. Recall from section 6.3 the expression for the time required by a bead to slide under gravity (and no friction) along the curve $y = y(x)$ from $(0,0)$ to (L, L):

$$J = \int_0^L \sqrt{\frac{1 + \left(\frac{dy}{dx}\right)^2}{2gy}}\, dx = \frac{1}{\sqrt{2g}} \int_0^L \sqrt{\frac{1 + (y')^2}{y}}\, dx.$$

Thus, ignoring the constant factor of $1/\sqrt{2g}$, we have

$$F = \left\{\frac{1 + (y')^2}{y}\right\}^{1/2}.$$

Unlike the previous example for the straight line, this F has an explicit dependence on y. This is a complication, but, on the other hand, notice as well that this F does not explicitly depend on x. In this case, then, we can use a "reduced" form of the Euler-Lagrange equation, derived in 1868 by the Italian mathematician Eugenio Beltrami (1835–1900):

$$if \ \frac{\partial F}{\partial x} = 0 \quad then \ F - y'\frac{\partial F}{\partial y'} = constant,$$

a result called *Beltrami's identity* (derived in appendix G).

Substituting the F for the descent-time integral into Beltrami's identity gives (with K some constant),

$$\left\{\frac{1+(y')^2}{y}\right\}^{1/2} - y' \cdot \frac{1}{2}\left\{\frac{1+(y')^2}{y}\right\}^{-1/2} \cdot \frac{2y'}{y} = K.$$

Or, with just a little bit of algebra,

$$y\left[1+(y')^2\right] = \frac{1}{K^2},$$

which, replacing the constant $1/K^2$ with C, becomes

$$y\left[1 + \left(\frac{dy}{dx}\right)^2\right] = C.$$

But this is precisely the differential equation for $y = y(x)$ that was derived in section 6.2 for the brachistochrone using Bernoulli's optical analogy. This means that, once again, the Euler-Lagrange equation has given us the correct answer.

6.6 Galileo's Hanging Chain

This chapter started with Galileo, and in this section we return to him once more. Imagine a given length of flexible, linked chain hanging under gravity from nails at each end, as shown in figure 6.14. A modern version of the hanging chain is a telephone wire or power transmission line hanging from adjacent support poles

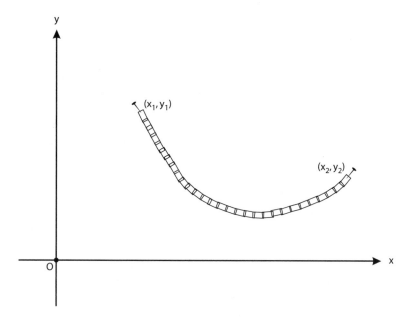

FIGURE 6.14. Galileo's hanging chain.

or towers. Our question here is easy to state: what is the shape of
the hanging chain? Galileo's answer, in his *Discourses* of 1638, was
equally plainspoken; a parabola. Since the Latin word for chain is
catena, the shape of the hanging chain was said (by Huygens in a
1690 letter to Leibniz) to be a *catenary*, and Galileo's claim, then,
was that the catenary is a parabolic curve.

Galileo was, however, wrong, and it was known for quite some
time that he was wrong. The German mathematician Joachim Jun-
gius (1587–1657) is generally given credit for formally establishing
this negative result in 1669. It wasn't until 1691, however, that
Leibniz and Johann Bernoulli actually figured out what the catenary
is. (You'll see, in just a bit, however, how it could be known that the
catenary is *not* a parabola—Huygens claimed to have known this
since his teenage years—without knowing what it actually is.)

We can solve Galileo's problem without the calculus of variations
and, of course, that is how it *must* have been solved in 1691, *five years*
before the brachistochrone challenge and long before the development of the Euler-Lagrange equation. Following how Bernoulli did
this will give us yet another solution that we can again compare to

what the Euler-Lagrange equation will tell us (and, of course, we *will* get the same answer). Bernoulli started by making the obvious observation that no matter how the chain hangs, it will have a lowest point. Let's call this point A, as shown in figure 6.15, and position the coordinate system so that A is on the y-axis. The tangent to the curve of the hanging chain at A is clearly horizontal. Bernoulli then said the chain to the left of $A(x < 0)$ is completely represented by the left-directed horizontal force it exerts on the rest of the chain to the right of A. Let's call that unknown force (or tension) at A, T_A; whatever it is, it is a constant.

Bernoulli then marked B as an arbitrary point with coordinates (x, y) on the chain, as shown in figure 6.15. If the mass M of the chain is uniformly distributed along its length L, then the mass density is simply $\rho = M/L$, a constant. So, if the length of the section of chain from A to B is s, and if as usual g is the acceleration of gravity, then the weight of the chain section between A and B is $\rho s g$. Finally, Bernoulli denoted the tension in the chain at B by T, directed of course along the tangent to the chain's curve at B. Call the angle that T makes at B, with the horizontal, θ (as shown in figure 6.15). Bernoulli then used the physical observation that the section of chain from between A and B is *not moving*. Thus, the net force acting on the chain must be zero. In particular, the sum of the

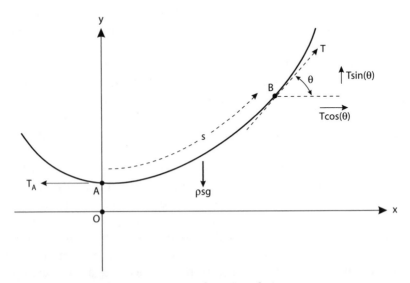

FIGURE 6.15. Static forces acting on a hanging chain.

horizontal forces, and the sum of the vertical forces, must *each* be zero. Thus,

$$T_A = T \cos(\theta), \qquad \text{(horizontal force equation)}$$

and

$$\rho gs = T \sin(\theta), \qquad \text{(vertical force equation)}.$$

Dividing these two equations into each other, and writing the constant $T_A/\rho g$ as simply k, we have

$$\frac{T \sin(\theta)}{T \cos(\theta)} = \frac{\rho gs}{T_A} = \tan(\theta) = \frac{1}{k}s.$$

Since $\tan(\theta)$ is the slope of the catenary at B, we can write

$$\frac{dy}{dx} = \frac{1}{k}s.$$

Differentiating both sides with respect to x gives Bernoulli's differential equation for the catenary:

$$\frac{d^2y}{dx^2} = \frac{1}{k}\frac{ds}{dx} = \frac{1}{k}\frac{\sqrt{(dx)^2 + (dy)^2}}{dx} = \frac{1}{k}\sqrt{1 + \left(\frac{dy}{dx}\right)^2}.$$

At this point I'll depart from Bernoulli's calculations (which became quite complicated) and continue with a pretty little trick that wasn't developed until years later (in 1712) but which will allow us to neatly and quickly solve Bernoulli's differential equation for $y = y(x)$. Following the lead of the Italian mathematician Jacopo Francesco Riccati (1676–1754), let's define the new variable p as

$$p = \frac{dy}{dx}, \qquad \text{and so} \quad \frac{dp}{dx} = \frac{d^2y}{dx^2}.$$

Then, Bernoulli's equation becomes

$$\frac{dp}{dx} = \frac{1}{k}\sqrt{1 + p^2}, \qquad \text{or } dx = \frac{k\, dp}{\sqrt{1 + p^2}},$$

which can now be easily integrated.

A table of indefinite integrals tells us, in fact, that

$$x = k \sinh^{-1}(p) + C_1$$

where C_1 is the constant of integration, and so

$$p = \sinh\left(\frac{x - C_1}{k}\right) = \frac{dy}{dx}.$$

This is even easier to integrate once more; since the derivative of the hyperbolic cosine is the hyperbolic sine, we can immediately write the equation of the catenary as

$$y(x) = k \cosh\left(\frac{x - C_1}{k}\right) + C_2,$$

where C_2 is another constant of integration.

We clearly have three adjustable constants to play with here (remember, k has the unknown tension T_A in it), and they can be determined from three conditions. Obvious candidates for those conditions are the coordinates of the ends of the catenary, and the length of the hanging chain. To be honest, however, the numerical work in finding those three constants can in general be a formidable task, and I will pursue it no further. (A short but quite interesting essay on how to do the number crunching is by Paul Cella, "Reexamining the Catenary," *College Mathematics Journal*, November 1999, pp. 391–93.)

I have written before of Johann Bernoulli's obsessively competitive personality, and the solution of the catenary problem provides yet another illustration of that unpleasant aspect of his nature. Writing a quarter of a century later of his memories of the competition between himself and his brother Jacob, he still gloried as much in his long-dead brother's failure as in his own success. As he wrote in a 1718 letter to a French correspondent:

> The efforts of my brother were without success; for my part, I was more fortunate, for I found the skill (I say it without boasting, why should I conceal the truth?) to solve it in full and to reduce it to the rectification of the parabola. It is true that it cost me study that robbed me of rest for an entire night. It was much

for those days and for the slight age and practice I then had, but the next morning, filled with joy, I ran to my brother, who was still struggling miserably with this Gordian knot without getting anywhere, always thinking like Galileo that the catenary was a parabola. Stop! Stop! I say to him, don't torture yourself any more to try to prove the identity of the catenary with the parabola, since it is entirely false. The parabola indeed serves in the construction of the catenary, but the two curves are so different that one is algebraic, the other is transcendental.

(This example of Johann's petty nastiness toward Jacob was not an isolated case. In an earlier 1712 letter to a mathematician in England, Johann called the forthcoming posthumous publication of Jacob's seminal probability book *Ars Conjectandi* "A monster which bears my brother's name.")

One last point. We see *now* that the catenary is not a parabola as Galileo believed because we have calculated what it actually is (a hyperbolic cosine). How could it have been discovered that the catenary is not a parabola *without* finding what it actually is? This can be done by a simple negative demonstration. That is, suppose we examine the curve of the suspension cable of a bridge, attached to a heavy bridge roadway, as shown in figure 6.16. Now we have a chain (the cable) hanging *not* by virtue of just its own mass, but also because of the enormously greater load of the massive bridge

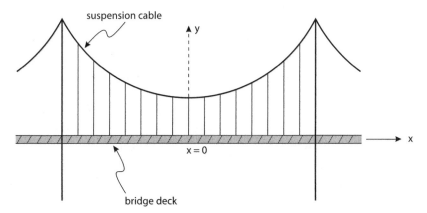

FIGURE 6.16. A hanging cable with uniform horizontal loading is a parabola.

deck connected to the suspension cable by a very large number of vertical hanger wires. In the original catenary problem, the force on the cable was due just to the mass of the cable itself, but in the suspension cable case, the cable mass is insignificant compared to the mass of the supported bridge deck. If we suppose the bridge deck has a uniform mass distribution along its length, and if we position the coordinate axes as shown in figure 6.16, then the equation for the catenary

$$\frac{dy}{dx} = \frac{1}{k}s$$

is replaced with

$$\frac{dy}{dx} = Kx,$$

where K is some constant. This is immediately integrable to give

$$y = \frac{1}{2}Kx^2 + C_2,$$

or, writing $\frac{1}{2}K = C_1$, we have the parabola

$$y = C_1 x^2 + C_2.$$

That is, assuming uniform mass density *along the cable* gives a catenary, while assuming uniform mass density *along the x-axis* gives the parabola. Two different assumptions, two different curves.

The physical interpretations of the constants C_1 and C_2 for the suspension bridge cable's parabolic curve are easy to see. C_2 is simply the height of the cable's low point (at $x = 0$) above the bridge deck. Also, if the tops of adjacent, uniform height support towers are at $(-a, b)$ and (a, b), where a and b are both positive, then $C_1 = (b - C_2)/a^2$.

As with the brachistochrone, the catenary has been mentioned in fictional literature. One character in Mark Helprin's 1983 novel *Winter's Tale*, for example, proclaims:

A bridge is a very special thing. Haven't you seen how delicate they are in relation to their size? They soar like birds; they extend and embody our finest efforts; and they utilize the curve of heaven. When a catenary of steel a mile long is hung in the clear over a river, believe me, God knows. . . . I would go as far as to say that the catenary, this marvelous graceful thing, this joy of physics, this perfect balance between rebellion and obedience, is God's own signature on earth. I think it pleases Him to see them raised.

Beautiful words, yes, but it would of course have been better for Helprin to have clearly distinguished between the unloaded hyperbolic catenary and the *loaded* parabola as the curve of the bridge's cable.

6.7   The Catenary Again

To see how the calculus of variations handles the catenary problem, let's return to the physical principle we used in the analysis of de L'Hospital's pulley problem in section 5.7. There it was argued that a massive body hanging under the effect of gravity alone (as does Galileo's chain—see figure 6.14 again) will hang in such a way as to minimize its total gravitational potential energy. For the pulley problem we needed only ordinary calculus, as the pulley was modeled as a single *point mass* and the supporting cables were taken as massless. For the catenary problem, however, the entire length of hanging chain has mass and so we have a massive, *spatially distributed* body.

To set the catenary problem in mathematics, let's assume as before that the chain's mass is uniformly distributed with constant density ρ. If we look at a differential length (ds) of the chain, located at the arbitrary point (x, y) on the curve $y = y(x)$, then the differential mass is $dm = \rho ds$ and the potential energy of that differential mass is $(\rho ds)gy$, where g is, as usual, the acceleration of gravity. Thus, the total potential energy of the hanging chain, which we wish to minimize by finding the "right" curve $y = y(x)$, is given by the integral

$$J = \int_{x_1}^{x_2} \rho g y \, ds = \int_{x_1}^{x_2} \rho g y \sqrt{(dx)^2 + (dy)^2}$$

$$= \int_{x_1}^{x_2} \rho g y \sqrt{1 + \left(\frac{dy}{dx}\right)^2} \, dx = \int_{x_1}^{x_2} \rho g y \sqrt{1 + (y')^2} \, dx.$$

Complicating matters just a bit here, however, is the first of the two extensions I mentioned back in section 6.4—we need to minimize J *under the constraint* of a given, fixed length of chain. That is, we must find that $y = y(x)$ that minimizes J while keeping the chain's length L constant:

$$L = \int_{x_1}^{x_2} ds = \int_{x_1}^{x_2} \sqrt{(dx)^2 + (dy)^2} = \int_{x_1}^{x_2} \sqrt{1 + (y')^2}\, dx = \text{constant.}$$

The clever idea that allows this additional twist to be taken into account is simply to write

$$\int_{x_1}^{x_2} \sqrt{1 + (y')^2}\, dx - L = 0 = \int_{x_1}^{x_2} \left\{ \sqrt{1 + (y')^2} - \frac{L}{x_2 - x_1} \right\} dx,$$

and then to argue that we can add *zero* (as many times as we wish) to J and we will have changed nothing! That is, with λ *anything* (λ is formally called a *Lagrange multiplier*) we will minimize not J but rather

$$\int_{x_1}^{x_2} \left[\rho g y \sqrt{1 + (y')^2} + \lambda \left\{ \sqrt{1 + (y')^2} - \left(\frac{L}{x_2 - x_1} \right) \right\} \right] dx.$$

The integrand function to be inserted into the Euler-Lagrange equation is therefore

$$F = \rho g y \sqrt{1 + (y')^2} + \lambda \sqrt{1 + (y')^2} - \frac{\lambda L}{x_2 - x_1}.$$

Notice, however, that the last term, if we take λ as any *constant*, is itself a constant and thus it will immediately vanish upon taking any derivative. So, the actual F that we need to consider is simply

$$F = \rho g y \sqrt{1 + (y')^2} + \lambda \sqrt{1 + (y')^2}.$$

That is, F is the integrand of the integral we wish to minimize *plus* a yet undetermined constant multiple of the constraint integral's integrand.

Notice that this F does not have an explicit dependence on x, and so we can use the already partially integrated form of the Euler-Lagrange equation called Beltrami's identity, just as we did in section 6.5. That is,

$$-F + y'\frac{\partial F}{\partial y'} = C_1 \qquad \text{(a constant)}.$$

Since

$$\frac{\partial F}{\partial y'} = \rho g y \frac{1}{2}\left\{1+(y')^2\right\}^{-1/2} 2y' + \lambda \frac{1}{2}\left\{1+(y')^2\right\}^{-1/2} 2y'$$

$$= \frac{\rho g y y'}{\sqrt{1+(y')^2}} + \frac{\lambda y'}{\sqrt{1+(y')^2}},$$

then we have

$$-\rho g y \sqrt{1+(y')^2} - \lambda\sqrt{1+(y')^2} + \frac{\rho g y (y')^2}{\sqrt{1+(y')^2}} + \frac{\lambda(y')^2}{\sqrt{1+(y')^2}} = C_1.$$

This reduces, after just a bit of simple algebra, to

$$(y')^2 = \frac{(\rho g y + \lambda)^2 - C_1^2}{C_1^2}.$$

We can now get two useful expressions from this. First,

$$\rho g y + \lambda = C_1\sqrt{1+(y')^2}.$$

And second, differentiation with respect to x of the $(y')^2$ expression gives

$$2y'y'' = \frac{2(\rho g y + \lambda)\,\rho g y'}{C_1^2},$$

which reduces (with the aid of the first expression) to

$$y'' = \frac{(\rho g y + \lambda)\rho g}{C_1^2} = \frac{\rho g C_1 \sqrt{1+(y')^2}}{C_1^2} = \frac{\rho g}{C_1}\sqrt{1+(y')^2}.$$

But this is precisely the same differential equation Bernoulli arrived at in the previous section (with $C_{1\rho g}$ replaced with k) by summing the horizontal and vertical forces on a section of the hanging chain. And we have already solved that (and this) equation using Riccati's trick of defining $p = y'$.

Well, all this is wonderful, you say, but so far all we have done is use the calculus of variations to solve problems *previously* solved by other means! Is this all we are going to get from the Euler-Lagrange equation? The answer is *no*, and in the final sections of this chapter I'll show you a couple of problems that the Euler-Lagrange formulation handles easily (including, at last, a proof of the isoperimetric theorem) that we have not found solutions to before.

Before leaving the catenary, however, let me tell you about one last wonderful property it has, one that, while known for centuries, seems not to be *well* known. Imagine that you want to build an arch (e.g., the entrance to a church) out of a nonorganic material (i.e., something that won't rot) that is very strong in compression but weak in tension. Such materials are brick, concrete, and stone, materials readily available to construction engineers for thousands of years. (Wood is very strong in both compression and tension, but it eventually decays.) The trick, then, to using bricks, concrete, and stone when building strong, *durable* structures is to avoid tension; in particular, we should construct our arch so that at every point there is only compression.

Now, think of the catenary, the curve of a chain hanging in complete repose. It is, at every point, in tension only, i.e., there clearly is no point where a hanging chain is in compression. This was apparently first pointed out in 1675 by Newton's contemporary (and sometimes rival) Robert Hooke (1635–1703). (After Hooke loudly claimed *he* was the true discoverer of the inverse square law of gravity, Newton deleted all mention of Hooke from the *Principia*. It didn't pay to irritate Newton!) Further, Hooke went on to observe, if the hanging catenary was "frozen in place" (e.g., glue the links of the flexible chain together) *and then inverted*, the resulting arch would be in *compression* only, and at no point would there be tension. Thus, an inverted catenary is the best (strongest) curve for a stone arch.

Hooke did not publish this result as a formal mathematical deduction (he was, in fact, not a very good mathematician), and it wasn't until considerably after Hooke's time that that was done. This is illustrated in a letter (dated December 23, 1788) written by Thomas Jefferson, in reply to a letter he had received a few months earlier. In his letter, Jefferson's correspondent had described his uncertainty in deciding between using a circular or a catenarian arch for the curve of the iron support tubes in the construction of a bridge. In his reply, Jefferson reports having just read a treatise on bridge arches, written by the Italian mathematician Lorenzo Mascheroni (1750–1800) which showed "every part of the Catenary is in perfect equilibrium." A modern example of an inverted catenary arch is the huge St. Louis Gateway Arch, made of stainless steel and standing 630 feet high. You can find much more on the use of the catenary in construction in the beautiful little book by Jacques Heyman, *The Stone Skeleton: Structural Engineering of Masonry Architecture* (Cambridge University Press 1995).

Here's a calculus of variations problem, of engineering importance, for you to try your hand at, and to which I don't think you can guess the answer by intuition. It occurs in the statistical theory of communication and information (and so electrical engineers are interested in it), but you don't have to know anything about those fields to do the math. The pure mathematical problem is simply this:

find the $y(x)$ that maximizes $J = -\int_{-\infty}^{\infty} y(x) \ln\{y(x)\}dx$, subject to the constraints $y(x) = 0$ unless $0 \le x \le M$ and $\int_{-\infty}^{\infty} y(x)\,dx = 1$, where M is a given positive constant.

For those who are curious, $y(x)$ is the probability density function of some nonnegative random variable, the J integral is the *entropy* (a measure of information) of that random variable, and M is the maximum value of that random variable. The constraint integral is simply the obvious statement that the total probability that the random variable has a value *somewhere* between minus infinity and plus infinity is one. But, as I said before, you don't really need to know any of this to solve the problem. Can you see why the solution $y(x)$ *maximizes J* (as opposed to minimizing it)? The solution is at the end of this chapter.

6.8   The Isoperimetric Problem, Solved (at last!)

The first complete proof of the ancient isoperimetric problem using the calculus of variations is due to the German mathematician Karl

Weierstrass (1815–97), dating from the period 1879–82. He never published that work, perhaps because of extremely poor health (after 1860 he could lecture only while sitting down as a student wrote the mathematics on a blackboard), but a record of it nevertheless survived. His students kept detailed notes of his lectures, and in 1927 his isoperimetric analyses were at last published.

To cast the isoperimetric problem into the generic form required by the Euler-Lagrange equation will throw yet a new twist at us. Recall the formula given in section 6.2 for the area enclosed by a closed, non-self-intersecting curve C that is traced out by a clockwise moving point in the time interval from 0 to T: if the parametric equations of C are $x = x(t)$ and $y = y(t)$, then

$$\text{area enclosed by } C = \frac{1}{2} \int_0^T \left\{ y(t)\frac{dx}{dt} - x(t)\frac{dy}{dt} \right\} dt.$$

We want to find the C that maximizes this integral, *given* a fixed prescribed perimeter. That is, we want to maximize the area integral subject to a perimeter *constraint,* just as we had a length constraint in the previous section in the hanging chain problem. A differential length of the curve C is, of course,

$$ds = \sqrt{(dx)^2 + (dy)^2} = \sqrt{\left(\frac{dx}{dt}\right)^2 + \left(\frac{dy}{dt}\right)^2} \, dt,$$

and so the total perimeter is (in terms of *time*) given by

$$\int ds = \int_0^T \sqrt{\left(\frac{dx}{dt}\right)^2 + \left(\frac{dy}{dt}\right)^2} \, dt.$$

With the area and perimeter integrals written out explicitly like this we can see the new complication—both x and y are functions of a new, *third* variable, *time,* which we did not have in our earlier work. The easiest way to see how to handle this new feature is to simply back up and rederive the Euler-Lagrange equation, taking time into consideration (as you'll soon see, we will actually get *two* Euler-Lagrange equations). So, just as we did before, let's assume $x = x(t)$ and $y = y(t)$ *are* the parametric equations for the solution curve C that we seek, and write perturbations around the solution as

$$X(t) = x(t) + \varepsilon\mu_1(t)$$

$$Y(t) = y(t) + \varepsilon\mu_2(t).$$

$\mu_1(t)$ and $\mu_2(t)$ are now two differentiable, independent, arbitrary functions, and ε is some constant. Our problem is to find the extrema of the functional

$$J(\varepsilon) = \int_{t_1}^{t_2} F\left\{t, \varepsilon, X(t), \dot{X}(t), Y(t), \dot{Y}(t)\right\} dt,$$

where I am using Newton's dot-notation for *time* derivatives (see section 4.4) to distinguish them from our earlier derivatives that were with respect to x, i.e.,

$$\dot{x}(t) = \frac{dx}{dt} \quad \text{and } \dot{Y}(t) = \frac{d}{dt} Y(t).$$

So,

$$\dot{X}(x) = \dot{x}(t) + \varepsilon\dot{\mu}_1(t)$$

$$\dot{Y}(t) = \dot{y}(t) + \varepsilon\dot{\mu}_2(t).$$

We can write the perimeter constraint integral as

$$P(\varepsilon) = \int_{t_1}^{t_2} G\left\{t, \varepsilon, X(t), \dot{X}(t), Y(t), \dot{Y}(t)\right\} dt,$$

which is a given constant. So, doing as we did in the previous section, let's form the sum integral $J + \lambda P$ (where λ, the Lagrange multiplier, is for now *any* constant); finding the curve C that gives the extrema to $J + \lambda P$ will find the curve that gives the extrema of J while also satisfying the perimeter constraint.

What we have, then, is

$$J(\varepsilon) + \lambda P(\varepsilon) = \int_{t_1}^{t_2} (F + \lambda G)\, dt,$$

where F is the integrand of the original J integral and G is the integrand of the constraint integral. To keep the notation from becoming "busy," let's call $F + \lambda G = H$. Then,

$$J(\varepsilon) + \lambda P(\varepsilon) = \int_{t_1}^{t_2} H\left\{t, \varepsilon, X(t), \dot{X}(t), Y(t), \dot{Y}(t)\right\}\, dt,$$

and we know that if $J(\varepsilon) + \lambda P(\varepsilon)$ has an extrema, it occurs at $\varepsilon = 0$ (because that's how we constructed things!). Therefore, just as before,

$$\frac{d}{d\varepsilon}\{J(\varepsilon) + \lambda P(\varepsilon)\}\Big|_{\varepsilon=0} = 0 = \int_{t_1}^{t_2} \frac{\partial H}{\partial \varepsilon}\, dt.$$

Writing $\partial H/\partial \varepsilon$ out in terms of the other variables, we have

$$\frac{\partial H}{\partial \varepsilon} = \frac{\partial H}{\partial X} \cdot \frac{\partial X}{\partial \varepsilon} + \frac{\partial H}{\partial \dot{X}} \cdot \frac{\partial \dot{X}}{\partial \varepsilon} + \frac{\partial H}{\partial Y} \cdot \frac{\partial Y}{\partial \varepsilon} + \frac{\partial H}{\partial \dot{Y}} \cdot \frac{\partial \dot{Y}}{\partial \varepsilon}$$

$$= \frac{\partial H}{\partial X}\mu_1 + \frac{\partial H}{\partial \dot{X}}\dot{\mu}_1 + \frac{\partial H}{\partial Y}\mu_2 + \frac{\partial H}{\partial \dot{Y}}\dot{\mu}_2.$$

Because setting $\varepsilon = 0$ is equivalent to setting $X(t)$ to $x(t)$ and $Y(t)$ to $y(t)$, we have

$$0 = \int_{t_1}^{t_2}\left\{\frac{\partial H}{\partial x}\mu_1 + \frac{\partial H}{\partial \dot{x}}\dot{\mu}_1 + \frac{\partial H}{\partial y}\mu_2 + \frac{\partial H}{\partial \dot{y}}\dot{\mu}_2\right\}\, dt.$$

Analogous to what we did in the original derivation of the Euler-Lagrange equation, let's assume that both $\mu_1(t)$ and $\mu_2(t)$ vanish at times $t = t_1$ and $t = t_2$ (the given start and stop times of our functional integral J). That is, $\mu_1(t_1) = \mu_1(t_2) = \mu_2(t_1) = \mu_2(t_2) = 0$. Now, remember that we are free to *separately* choose $\mu_1(t)$ and $\mu_2(t)$ in any way we wish, as long as *both* vanish at $t = t_1$ and $t = t_2$. In particular, we could choose $\mu_2(t) = 0$ for *all* t. Then,

$$\int_{t_1}^{t_2}\left\{\frac{\partial H}{\partial x}\mu_1 + \frac{\partial H}{\partial \dot{x}}\dot{\mu}_1\right\}\, dt = 0.$$

Or we could choose $\mu_1(t) = 0$ for *all* t, and thus conclude that

$$\int_{t_1}^{t_2}\left\{\frac{\partial H}{\partial y}\mu_2 + \frac{\partial H}{\partial \dot{y}}\dot{\mu}_2\right\}\, dt = 0.$$

Just as in our earlier derivation of the Euler-Lagrange equation, we now simply integrate (by parts) the second term of each of these two integrals. Everything goes through just as before (I'll leave the filling in of all the easy details for you), but now we will end up with *two* Euler-Lagrange equations. I'll enclose them in a box because of their central importance in the calculus of variations:

$$\frac{\partial H}{\partial x} - \frac{d}{dt}\left(\frac{\partial H}{\partial \dot{x}}\right) = 0$$

$$\frac{\partial H}{\partial y} - \frac{d}{dt}\left(\frac{\partial H}{\partial \dot{y}}\right) = 0$$

Now, *at last*, we can finally solve the isoperimetric problem. We have, from the start of this section, the integrand of the area functional as

$$F = \frac{1}{2}(y\dot{x} - x\dot{y})$$

and the integrand of the perimeter constraint as

$$G = \left(\dot{x}^2 + \dot{y}^2\right)^{1/2}.$$

Thus,

$$H = \frac{1}{2}(y\dot{x} - x\dot{y}) + \lambda\left(\dot{x}^2 + \dot{y}^2\right)^{1/2}.$$

So, for our first Euler-Lagrange equation, we have

$$\frac{\partial H}{\partial \dot{x}} = \frac{1}{2}y + \frac{1}{2}\lambda\left(\dot{x}^2 + \dot{y}^2\right)^{-1/2} 2\dot{x}$$

$$= \frac{1}{2}y + \lambda\dot{x}\left(\dot{x}^2 + \dot{y}^2\right)^{-1/2},$$

and therefore

$$\frac{d}{dt}\left(\frac{\partial H}{\partial \dot{x}}\right) = \frac{1}{2}\dot{y} + \frac{d}{dt}\left[\frac{\lambda\dot{x}}{\sqrt{\dot{x}^2 + \dot{y}^2}}\right].$$

Since

$$\frac{\partial H}{\partial x} = -\frac{1}{2}\dot{y},$$

then

$$-\frac{1}{2}\dot{y} - \frac{1}{2}\dot{y} - \frac{d}{dt}\left[\frac{\lambda \dot{x}}{\sqrt{\dot{x}^2 + \dot{y}^2}}\right] = 0,$$

or

$$\frac{d}{dt}\left[y + \frac{\lambda \dot{x}}{\sqrt{\dot{x}^2 + \dot{y}^2}}\right] = 0,$$

which can be integrated by inspection to give

$$y + \frac{\lambda \dot{x}}{\sqrt{\dot{x}^2 + \dot{y}^2}} = C_1, \qquad \text{a constant.}$$

Thus, at last,

$$y - C_1 = -\frac{\lambda \dot{x}}{\sqrt{\dot{x}^2 + \dot{y}^2}}.$$

Repeating the calculations for the second Euler-Lagrange equation, we have

$$\frac{\partial H}{\partial \dot{y}} = -\frac{1}{2}x + \frac{1}{2}\lambda\left(\dot{x}^2 + \dot{y}^2\right)^{-1/2} 2\dot{y}$$

$$= -\frac{1}{2}x + \lambda\dot{y}\left(\dot{x}^2 + \dot{y}^2\right)^{-1/2},$$

and so

$$\frac{d}{dt}\left(\frac{\partial H}{\partial \dot{y}}\right) = -\frac{1}{2}\dot{x} + \frac{d}{dt}\left[\frac{\lambda \dot{y}}{\sqrt{\dot{x}^2 + \dot{y}^2}}\right].$$

Since

$$\frac{\partial H}{\partial y} = \frac{1}{2}\dot{x},$$

then

$$\frac{1}{2}\dot{x} + \frac{1}{2}\dot{x} - \frac{d}{dt}\left[\frac{\lambda \dot{y}}{\sqrt{\dot{x}^2 + \dot{y}^2}}\right] = 0,$$

or

$$\frac{d}{dt}\left[x - \frac{\lambda \dot{y}}{\sqrt{\dot{x}^2 + \dot{y}^2}} \right] = 0,$$

which immediately integrates to

$$x - \frac{\lambda \dot{y}}{\sqrt{\dot{x}^2 + \dot{y}^2}} = C_2, \quad \text{a constant.}$$

And so, at last,

$$x - C_2 = \frac{\lambda \dot{y}}{\sqrt{\dot{x}^2 + \dot{y}^2}}.$$

Squaring and adding the $(x - C_2)$ and $(y - C_1)$ expressions gives

$$(y - C_1)^2 + (x - C_2)^2 = \frac{\lambda^2 \dot{x}^2 + \lambda^2 \dot{y}^2}{\dot{x}^2 + \dot{y}^2} = \lambda^2.$$

But this is just the equation for a *circle* (yes!) of radius λ centered on the point (C_2, C_1). We can, of course, pick the integration constants C_1 and C_2 to center the circle anywhere we wish. We now see, too, that if P is the given perimeter of the closed curve of maximum area, then $2\pi\lambda = P$, i.e., the Lagrange multiplier constant (of previously unknown value) is equal to $P/2\pi$. Once again the Euler-Lagrange formulation has formally given us what we "knew" to be the answer to an historically important problem. But now we have a *proof*, and that is what a mathematician wants. Intuition, after all, is too often a passport to the "land of error"!

Now that the mathematical certainty of the isoperimetric theorem has been established, let me end this section with a challenge problem for you. Unlike the other challenges in this book, this one does not come with an answer, because I don't have one. And yet, it should require only first-year calculus. To start, consider figure 6.17, which shows an ellipse (divided into four quarters) with semi-major axes of lengths a and b. The area of this ellipse is (a standard freshman calculus problem) given by $\pi a b$. In figure 6.18, the four quarters have been rearranged to form a new figure with area $\pi a b + (a - b)^2$. The crucial observation about these two figures is that they have the *same perimeter* (I'll call it P).

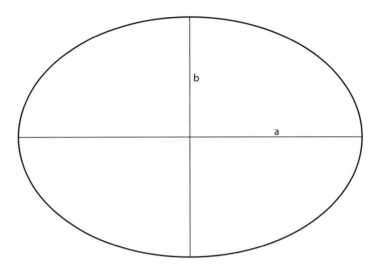

FIGURE 6.17. An ellipse.

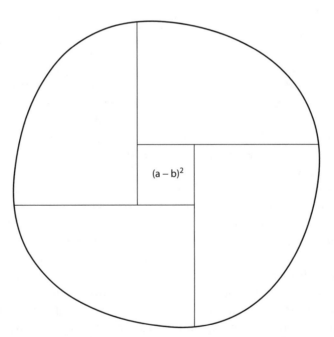

FIGURE 6.18. The ellipse of figure 6.17 quartered and rearranged (same perimeter but increased area).

The parametric equations for the ellipse are

$$x(t) = -a \cos(t)$$

$$y(t) = b \sin(t),$$

equations that describe a point traveling in a clockwise sense along the boundary edge of the ellipse. The point makes one complete orbit of the perimeter in time interval 2π (see appendix F). From the formula for the length of a curve (see section 6.3), we then have

$$P = \int_0^{2\pi} \sqrt{(dx)^2 + (dy)^2} = \int_0^{2\pi} \sqrt{\left(\frac{dx}{dt}\right)^2 + \left(\frac{dy}{dt}\right)^2} \, dt$$

$$= \int_0^{2\pi} \sqrt{a^2 \sin^2(t) + b^2 \cos^2(t)} \, dt.$$

Now, the isoperimetric theorem says that the area of a plane region with a perimeter of $P = 2\pi R$ cannot exceed the area of a circle with radius R. That is,

$$A \leq \pi R^2 = \pi \left(\frac{P}{2\pi}\right)^2 = \frac{P^2}{4\pi}.$$

Thus, $P \geq \sqrt{4\pi A}$, and so, using the area of figure 6.18 for A, we have

$$\int_0^{2\pi} \sqrt{a^2 \sin^2(t) + b^2 \cos^2(t)} \, dt \geq \sqrt{4\pi \left\{\pi ab + (a - b)^2\right\}},$$

where the equality obviously holds when $a = b$.

Here's the challenge—there seems to be no "easy" way to derive this inequality *directly*, by manipulating the integral on the left-hand side. That is, *I* can't see how to do it. If you try your hand at it and succeed, please write to me and tell me how you did it!

6.9   Minimal Area Surfaces, Plateau's Problem, and Soap Bubbles

In 1744, Euler solved the following purely mathematical problem, and thereby started an area of research that continues to this day:

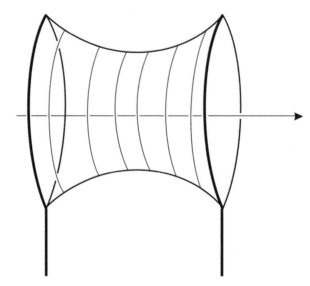

Figure 6.19. Surface of revolution with circular ends.

"If we connect the two given points (x_1, y_1) and (x_2, y_2) with a curve $y = y(x) \geq 0$, and then revolve that curve about the x-axis, a 'cylinder-like' surface will be created (with circular, *open* ends). What should that curve be to make the area of the surface as small as possible?" Euler actually considered the slightly less general case of $y_1 = y_2$ (the open circular ends have the same radius), but the case of $y_1 \neq y_2$ offers no additional complications and so that's the problem I'll discuss in this section, with reference to figure 6.19.

With ds as a differential length along the curve $y = y(x)$, then the differential surface area dA swept out by revolving ds about the x-axis is

$$dA = (2\pi y)ds = 2\pi y\sqrt{(dx)^2 + (dy)^2} = 2\pi y\sqrt{1 + (y')^2}\, dx,$$

and so the total area of the "cylinder-like" surface is

$$J = \int dA = 2\pi \int_{x_1}^{x_2} y\sqrt{1 + (y')^2}\, dx.$$

Thus, the F to be inserted into the Euler-Lagrange equation of section 6.4 is simply (ignoring the constant factor of 2π)

$$F = y \left\{ 1 + (y')^2 \right\}^{1/2}.$$

Since F has no explicit dependence on x, we can use Beltrami's partially integrated form of the Euler-Lagrange equation,

$$F - y' \frac{\partial F}{\partial y'} = \text{constant} = C_1.$$

Now

$$\frac{\partial F}{\partial y'} = y \frac{1}{2} \left\{ 1 + (y')^2 \right\}^{-1/2} 2y' = \frac{yy'}{\sqrt{1 + (y')^2}},$$

and so Beltrami's identity becomes

$$y\sqrt{1 + (y')^2} - \frac{y(y')^2}{\sqrt{1 + (y')^2}} = C_1,$$

or

$$\boxed{\frac{y}{\sqrt{1 + (y')^2}} = C_1.}$$

I've put this result in a box as I'll be referring back to it soon.

Thus, if we square and multiply, we have

$$y^2 = C_1^2 + C_1^2 \, (y')^2,$$

and differentiation with respect to y gives

$$2yy' = 2C_1^2 \, y'y'',$$

or, using the boxed result,

$$\frac{y''}{y} = \frac{1}{C_1^2} = \frac{y''}{C_1\sqrt{1 + (y')^2}},$$

which becomes

$$y'' = \frac{1}{C_1}\sqrt{1 + (y')^2}.$$

But this is just Bernoulli's equation (with k written instead of C_1) for the catenary, derived in section 6.6! The surface formed by rotating a catenary is called a *catenoid*, and so the answer to Euler's surface problem is seen to be intimately related to Galileo's problem of the hanging chain. Amazing! But this is only the *start* of the amazing results that flow from Euler's pioneering calculation.

A fascinating physical interpretation of Euler's problem is found in the physics of soap films. Such films (or bubbles) are easily made from ordinary dishwashing detergent, warm water and, if desired, some glycerin to add stability to the films. Soap films have the property that their surface energy is proportional to their surface area, which means a minimum *energy* film (what Nature "strives" for) is equivalent to a minimum *area* film. [For a brief but quite interesting physics tutorial on this point, see A. Fomenko, "Minimal Surfaces" (*Quantum*, May/June 2000, pp. 4–7, 13), as well as *the* classic paper by two (husband and wife) mathematicians: Frederick J. Almgren, Jr., and Jean E. Taylor, "The Geometry of Soap Films and Soap Bubbles" (*Scientific American*, July 1976, pp. 82–93).] Therefore, to *experimentally* solve Euler's problem all one need do is dip two circular wire rings into a soap solution and allow a film to form between them, as shown in figure 6.19.

Euler's problem is, in fact, a special case of the so-called *Plateau problem*: given a contour in space, show that a surface of minimal area bounded by that contour exists. The name comes from the Belgium physicist Joseph Plateau (1801–83) who, over the period 1843–69, experimentally studied minimal areas using wire contours dipped into soap solution (more on Plateau is in the final section of this chapter). The problem actually dates, however, from about 1761, when it was posed by Lagrange. Lagrange's formulation of the Plateau problem asks for the demonstration of a surface of minimum area for any given *single* contour edge, i.e., for any given frame consisting of a single closed length of wire. Note carefully that the two unconnected circular rings of Euler's problem are *not* such a contour, and that if the centers of the two rings are sufficiently displaced, laterally, then a soap film will *not* form. More precisely, there is no minimal surface for Euler's two-ring contour if the projections of the two rings do not have some overlap [see Johannes C. C. Nitsche, "Plateau's Problems and Their Modern Ramifications" (*American Mathematical Monthly*, November 1974, pp. 945–68)].

For a *single*, closed contour, however, no matter how bizarre its twists and turns in three-dimensional space may be, the answer to Lagrange's original question is *yes*, there is *always* a minimal surface. That was first proven in 1931 by the American mathematician Jesse Douglas (1897–1965) and, independently in 1933, by the Hungarian-born American mathematician Tibor Radó (1895–1965). What Douglas and Radó provided were existence proofs, which is, of course, not the same thing as actually displaying the specific minimal surface that goes with a given closed contour as its edge. Specific minimal surfaces for given edges are generally quite difficult to find; for example, in 1890, H. A. Schwarz (see the box at the end of section 2.6) found the minimal surface determined by a skew quadrilateral contour, as shown in figure 6.20. Nitsche's paper, cited above, gives

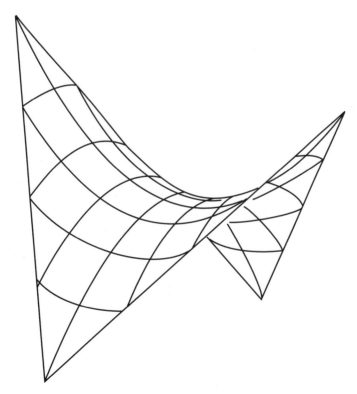

FIGURE 6.20. Surface with a skew quadrilateral boundary.

the solution—its expression requires three hyperelliptic integrals, i.e., it is *complicated!*

The general solution for Euler's $y = y(x)$ curve was shown in section 6.6 to be the hyperbolic cosine, i.e., with C_1 and C_2 as adjustable constants (and now no length constraint),

$$y = C_1 \cosh \left(\frac{x - C_2}{C_1} \right).$$

To keep things mathematically nice, let's now drop back down to Euler's original problem with $y_1 = y_2$. In particular, let's write $y_1 = y_2 = y_0 = 1$, and position the y-axis so that $x_1 = -x_0 < 0$ and $x_2 = x_0 > 0$. That is, the two soap rings are $2x_0$ apart. By symmetry we have y minimum at $x = 0$, and so $C_2 = 0$. Thus,

$$y = C_1 \cosh \left(\frac{x}{C_1} \right).$$

To be even more particular (so we can calculate some numbers), suppose $x_0 = \frac{1}{2}$. Then we have $y = 1$ at both ends $\left(x = \pm \frac{1}{2} \right)$, and so

$$1 = C_1 \cosh \left(\frac{1}{2C_1} \right).$$

This is a transcendental equation for C_1, which means we can't solve explicitly in closed form for C_1. We need to resort to numerical methods to find the value of C_1 and this is, in fact, a perfect problem to which we can apply the Newton-Raphson algorithm developed in section 4.5. A plot of $f(C_1) = C_1 \cosh(1/2C_1) - 1$ shows (see figure 6.21) that there are actually two values of C_1 that satisfy $f(C_1) = 0$. This may at first seem puzzling as, after all, the minimal area surface would seem to be a *unique* surface. Do two solutions to $f(C_1) = 0$ mean that a soap film can be either one of *two* possible shapes? The answer is *no*, and this will be explained by the end of this section. For now, however, the plot in figure 6.21 gives us initial guesses with which to start the Newton-Raphson algorithm, which fine-tunes the solutions to $f(C_1) = 0$ when $C_1 = 0.235$ and $C_1 = 0.848$.

If we separate the two rings even more, from $x_0 = \frac{1}{2}$ to $x_0 = 1$, then we have yet another surprise waiting for us. With $x_0 = 1$, our condition on C_1 becomes (at the rings where $y = y_0 = 1$)

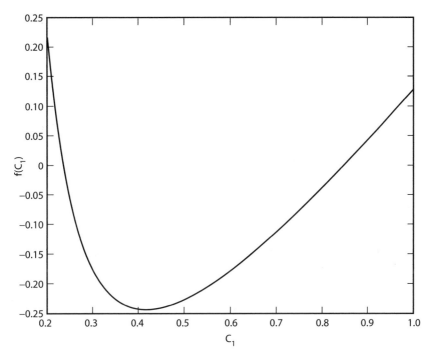

FIGURE 6.21. $f(C_1) = C_1 \cosh(1/2C_1) - 1$.

$$1 = C_1 \cosh\left(\frac{1}{C_1}\right).$$

As figure 6.22 shows, there is now *no* real solution for C_1, i.e., the plot of $f(C_1) = C_1 \cosh(1/C_1) - 1$ never crosses the C_1 axis! We can understand what this means, *mathematically*, as follows. As x_0 increases from $\frac{1}{2}$ to 1, the $f(C_1)$ curve "rises upward" and, at some critical value of x_0 (call it $\hat{x}_0$), the two crossings of the C_1 axis merge together into a double root. For $x_0 > \hat{x}_0$ the $f(C_1)$ curve rises above the C_1 axis and there are no crossings (no real solutions). What is happening, *physically*, as we increase x_0, is that at $x_0 = \hat{x}_0$ the soap film *breaks*, and for $x_0 > \hat{x}_0$ there is *no* cylindrical soap film surface connecting the two rings.

So, what happens to the soap film after it breaks when x_0 exceeds $\hat{x}_0$? The answer (verified experimentally) is that it forms two circular films, one at each ring. Reducing x_0 to less than $\hat{x}_0$ does *not*, of course,

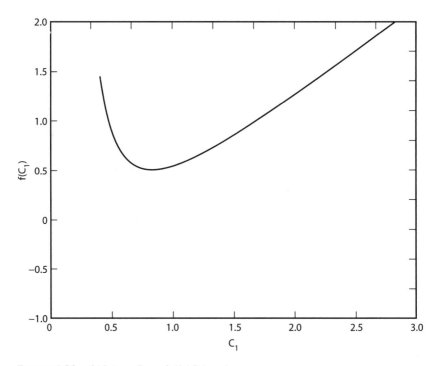

FIGURE 6.22. $f(C_1) = C_1 \cosh(1/C_1) - 1$.

cause the catenoid surface to reappear, and so the breaking of the soap film is both sudden and irreversible. This discontinuous behavior is called the *Goldschmidt solution*, after the German mathematician C.W.B. Goldschmidt (1807–51) who discovered it (on paper) in 1831.

We can calculate the value of $\hat{x}_0$, the maximum value of x_0 that can support a catenoid minimum area soap film surface in Euler's problem, as follows. We have, from before, that at the circular rings (where $x = \pm x_0$ and $y = 1$),

$$\frac{1}{C_1} = \cosh\left(\frac{x_0}{C_1}\right).$$

We also know from our earlier discussion that a plot (for a given x_0) of $f(C_1) = \cosh(x_0/C_1) - (1/C_1)$ will, in general, have *two* solutions to $f(C_1) = 0$. When $x_0 = \hat{x}_0$, however, the plot of $f(C_1)$ will just touch the C_1 axis at a single point. This means the C_1 axis is tangent to the $f(C_1)$ curve when $x_0 = \hat{x}_0$, and so

$$\frac{df}{dC_1}\bigg|_{x_0=\hat{x}_0} = 0.$$

So,

$$\frac{df}{dC_1} = -\frac{x_0}{C_1^2}\sinh\left(\frac{x_0}{C_1}\right) + \frac{1}{C_1^2},$$

and thus, when $x_0 = \hat{x}_0$,

$$0 = -\hat{x}_0\sinh\left(\frac{\hat{x}_0}{C_1}\right) + 1,$$

or

$$\hat{x}_0\sinh\left(\frac{\hat{x}_0}{C_1}\right) = 1.$$

We also have, of course, that

$$\cosh\left(\frac{\hat{x}_0}{C_1}\right) = \frac{1}{C_1}.$$

Dividing these two results into each other gives

$$\frac{\cosh\left(\dfrac{\hat{x}_0}{C_1}\right)}{\hat{x}_0\sinh\left(\dfrac{\hat{x}_0}{C_1}\right)} = \frac{1}{C_1} = \frac{1}{\hat{x}_0}\coth\left(\frac{\hat{x}_0}{C_1}\right),$$

or

$$\coth\left(\frac{\hat{x}_0}{C_1}\right) = \frac{\hat{x}_0}{C_1}.$$

This can be solved (numerically) to give $\hat{x}_0/C_1 = 1.1997$, and so

$$\hat{x}_0 = \frac{1}{\sinh\left(\dfrac{\hat{x}_0}{C_1}\right)} = \frac{1}{\sinh(1.1997)} = 0.6627.$$

In summary, if we have two wire rings each of unit radius, then there is a catenoid soap film if their separation is less than $2\hat{x}_0 = 1.3254$, and there can *not* be such a surface if their separation is greater than 1.3254.

Finally, to clean up the last loose end of this section, we need to explain why (for a given ring separation of $2x_0$) there are generally *two* possible values for C_1. To understand this, let's calculate the actual surface area of the soap film catenoid as a function of C_1. We have, by symmetry, that this area is simply twice the area of half of the catenoid surface, i.e., of the surface from one end to halfway to the other end:

$$A = 2(2\pi) \int_0^{x_0} y\sqrt{1 + (y')^2} \, dx,$$

where

$$y = C_1 \cosh\left(\frac{x}{C_1}\right), \qquad -x_0 \le x \le x_0.$$

But, looking back at the result (which I placed in a box at the beginning of this section) that we got from Beltrami's identity, we have

$$\sqrt{1 + (y')^2} = \frac{y}{C_1}.$$

Thus,

$$A = 4\pi \int_0^{x_0} \frac{y^2}{C_1} \, dx = 4\pi \, C_1 \int_0^{x_0} \cosh^2\left(\frac{x}{C_1}\right) dx.$$

From any good table of integrals, we find that

$$\int \cosh^2(u) \, du = \frac{\sinh(2u)}{4} + \frac{u}{2},$$

from which it immediately follows that

$$A = \pi \, C_1^2 \left[\sinh\left(\frac{2x_0}{C_1}\right) + \frac{2x_0}{C_1} \right].$$

For the case of $x_0 = \frac{1}{2}$, for example, we found earlier that $C_1 = 0.235$ or $C_1 = 0.848$. Evaluating A for each value, we find $A(C_1 = 0.235) =$

6.85 and $A(C_1 = 0.848) = 5.99$. Thus, $C_1 = 0.848$ is *the* value to use to have *the* minimal area catenoid, the one that is actually observed to form.

The discussion in this section on the Plateau problem of minimal area surfaces with a specified boundary edge has not even been a minimal scratch on the topmost surface of the topic (please forgive the outrageous pun!). Ever since Plateau's pioneering soap film studies, there have been more questions than answers, and minimal surfaces will surely be an active area of mathematical research for many decades to come. Two of the most fundamental questions *have*, however, only recently been answered: (1) the *reasons* for the empirical Plateau rules for how soap films connect to each other, and (2) the wonderfully named *double bubble conjecture*. Each is easy to understand, but each required deep mathematical attacks for their solution. I'll end this section with a paragraph on each.

His extensive examination of countless soap films led Plateau to the conclusion that those structures do not assume their shapes at random. Rather, they follow two simple rules. Either

1. three film surfaces connect along a common edge, with the surfaces making 120° angles with each other, or
2. six film surfaces connect at a common point (making four edges together) with an angle of about 109° between any two of the edges (the exact value is $\cos^{-1}\left(-\frac{1}{3}\right) = 109.47122\ldots°$).

Both of these rules are illustrated in figure 6.23, which shows the soap film that forms on a cubical wire frame (there are a total of 13 films meeting along common edges and/or points). These rules appear to explain every soap film ever observed, but that's hardly a proof that they actually do. There was always the possibility that a sufficiently complicated wire frame might result in a film structure not explainable by Plateau's rules alone. Only in 1976 (as a follow up to her 1972 Princeton doctoral dissertation) was it at last *proven* by the American mathematician Jean Taylor (1944–) that the rules follow as necessary *and sufficient* consequences of the surface-energy-minimizing property of soap films. You can find more on Plateau's rules, and their implications, in the following two papers: Cyril Isenberg, "Problem-Solving with Soap Films" (*Physics Teacher*, January 1977, pp. 9–18), and Dale T. Hoffman, "Smart Bubbles Can Do Calculus" (*Mathematics Teacher*, May 1979, pp. 377–88).

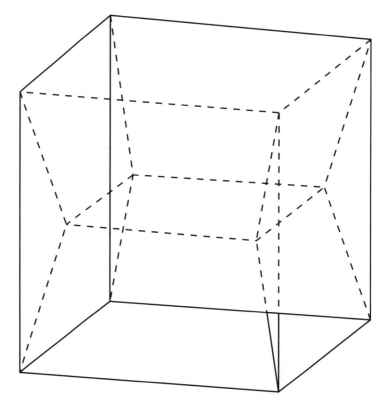

FIGURE 6.23. Plateau's rules illustrated on a cubical frame.

The double bubble conjecture says that if two prescribed but separate volumes are to be enclosed by the minimum surface area, then two bubbles made of three portions of spherical surfaces (one in common, of course) is the way to do it. Many have thought the double bubble conjecture to be obvious, e.g., the classic, popular 1890 book on soap bubbles is by the brilliant English experimentalist C. V. Boys (1855–1945)—*Soap Bubbles and the Forces Which Mould Them*—who wrote there of the spherical double bubble not as conjecture but as obvious fact. Boys was wrong, however, and while the conjecture is true it is not at all obvious. As astonishing as it may seem, just the *two*-dimensional version (substitute *areas* for *volumes*, and *perimeter* for *surface area*, and then figure 6.24 shows a *planar* double bubble for two equal prescribed areas) remained unproven until 1993. The three-dimensional case, for two volumes, was even

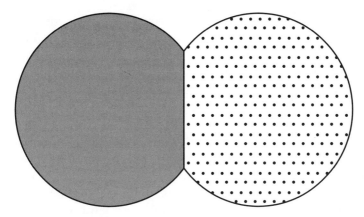

FIGURE 6.24. Double (planar) bubble.

tougher, resisting all efforts until 2000. You can read about these proofs (based in large part on the efforts of a team of *undergraduate* college mathematics students!) in the following two papers: Joel Foisy et al., "The Standard Double Soap Bubble in $\mathbf{R}^2$ Uniquely Minimizes Perimeter" (*Pacific Journal of Mathematics*, May 1993, pp. 47–58), and Frank Morgan, "Proof of the Double Bubble Conjecture" (*American Mathematical Monthly*, March 2001, pp. 193–205).

6.10   The Human Side of Minimal Area Surfaces

This last section to chapter 6 is a bit different from the rest of the book. It is all prose, with not a single equation. The reason is that, as I wrote the previous section on minimal surfaces and soap films, I came across some curious and (in one case) interlocking stories of the people (all mentioned in the last section) who did the pioneering mathematical and physical research. I wasn't able to weave any of that material into the mathematical discussions, but instead have saved these little vignettes for a section of their own. I'll start with C.W.B. Goldschmidt, the man who discovered the mathematics behind the breaking of the soap film solution to Euler's minimal surface problem.

Almost all books on the calculus of variations discuss the Goldschmidt solution, but none (as far as I know) says anything about the man. My curiosity was sparked by the silence, and so I searched for more information. That search eventually led to the discovery of a brief obituary notice that appeared in the 1851 volume of the *American Journal of Science* (pp. 443–44). There it was reported that Carl Wolfgang Benjamin Goldschmidt was a professor of astronomy at the University of Göttingen ("though perhaps not a great astronomer he was an enthusiastic and laborious one") and served as an assistant to the great Gauss at the observatory there. The notice made the observation, now ironic considering how history turned out, that "Goldschmidt's name will be long honored by those who never knew him." After mentioning "his investigation of the minimum surfaces of rotation of curves about a fixed axis," the notice ends by revealing (with a typically Victorian romantic view of death) the nature of Goldschmidt's early demise: "His death was like his life—quiet and peaceful. He had long suffered from the consequences of an enlargement of the heart; and on the morning of Feb. 15th, he was found in his bed, sleeping the sleep that knows no waking."

Just two years before Goldschmidt's insight into the Euler problem, the soap film pioneer Joseph Plateau conducted a fateful experiment that would lead to his loss of sight. At the University of Liège, while conducting his doctoral dissertation research in physiological optics (in particular, the formation of images on the retina), Plateau stared at the sun for nearly half a minute. How he managed to do this incredibly stupid thing without being under the influence of brain-deadening drugs has always mystified me, but he did. He ended up paying a very big price for his diploma—after temporarily losing his vision and then partially regaining it, by 1841 his corneas were severely inflamed and by 1843 he was completely and irreversibly blind. A very bad state for anyone, of course, and extraordinarily bad for an experimentalist. Or so one would think. His classic soap film experiments were just beginning, and so Plateau enlisted the eyes and help of his colleagues and students (at the University of Ghent) to make the actual observations from which were deduced "Plateau's rules." The laws governing one of Nature's most beautiful displays in the everyday world are, then, due to a blind man, an achievement that brings to mind the creation of beautiful music by the deaf Beethoven.

In 1855, even as Plateau was pondering the soap films he could no longer see, the man who would "popularize" them was born.

Charles Vernon Boys eventually became famous as an inspired experimental physicist, remembered to this day as the inventor of (among many inventions) the *Boys camera*. With that camera he was able to photograph rifle bullets in flight (at 1,400 miles per hour!), along with the acoustic shock waves they produced. You can find in-flight bullet images produced by Boys with his nineteenth-century camera, still fascinating to view in the twenty-first century, in *Nature* (March 2, pp. 415–21, and March 9, pp. 440–46, 1893). With his superfast camera you could even record the very bursting of a soap bubble by a bullet.

During the Christmas season of 1889–90, Boys delivered a series of lectures to a juvenile audience at the Royal Institution (London), and those lectures and the accompanying lantern slides were brought out as his famous book *Soap Bubbles and the Forces Which Mould Them*. The lectures (and book) were enormous successes, and both displayed wonderful teaching and expository skills. Interestingly, that wasn't always the case for Boys. In his 1934 *Experiment in Autobiography*, H. G. Wells revealed that he had been a former student of Boys in 1886, at the Normal School of Science (London), and that he had been singularly unimpressed. As Wells wrote of Boys, he was "[T]hen an extremely blond and largely inaudible young man already famous for his manipulative skill and ingenuity with soap bubbles. . . . In those days I thought him one of the worst teachers who has ever turned his back upon a resistive audience, messed about with the blackboard, galloped through an hour of talk and bolted back to the apparatus in his private room. . . . Boys was too fast." By the time Wells wrote those words he was world famous as the author of *The Time Machine*, *War of the Worlds*, and other "scientific romances," as he called his science fiction novels. Indeed, Wells was far more famous than was the still-living Boys. It would be interesting to know what Boys thought when he read Wells' description of him (and it is hard to imagine that it wasn't brought to his attention by someone).

Brilliant at experiment as he was, Boys had a dark side to him, too; he loved to play practical jokes on people, a sophomoric activity that mostly amuses the jokester. Certainly his wife, Marion, was not amused by her husband's antics; she put up with them from the start of their marriage in 1892, but finally divorced him in 1910. There is some speculation, even today, that Boys' treatment of his wife might have strayed as far south as to be labeled abusive, but in any case the marriage was so wounded that even before 1910 Marion had begun an affair with the Cambridge mathematician Andrew Forsyth (1858–1942). The two married after her divorce, but the resulting scandal forced Forsyth to resign from Cambridge. As his obituary notice in *Nature* put it, "In painful circumstances he made a marriage of affection, and gained ten years of a happiness for which he counted the loss of many old associations a price not too high." Forsyth later (1927), after Marion's death, published the influential book *Calculus of Variations* (its dedication is simply "To Marion in Remembrance") but in his discussion of Euler's minimal surface problem there is no mention, none at all, of soap films, bubbles, or Boys.

And finally, the authors of the *Scientific American* article on soap bubbles that is almost universally cited by authors writing on minimal surfaces are, in a sense, a mirror image reflection of Marion and Andrew. Jean Taylor was Frederick Almgren's (1933–97) first doctoral student at Princeton, and through him was introduced to minimal surfaces. (Her undergraduate degree, and a master's too, are in chemistry, representing scientific knowledge that certainly must have given her physical insight into the behavior of soap films that a strictly pure mathematician would lack.) Taylor and Almgren later married and continued their mutual work on minimal surfaces, work that eventually led to Taylor's solution to the century-old problem of explaining Plateau's rules. Jean Taylor (who I suspect was the inspiration for Rebecca Goldstein's fictional Princeton math professor Phoebe Saunders, an expert in the mathematics of soap films—see *Strange Attractors*) is presently professor of mathematics at Rutgers University.

Solution to the Problem in Section 6.7

Writing λ as the arbitrary, constant Lagrange multiplier that allows us to apply the constraints, we wish to minimize the integral

$$\int_{-\infty}^{\infty} \{-y\ln(y) + \lambda y\}\, dx.$$

(continued)

So,

$$F = -y \ln(y) + \lambda y,$$

and thus

$$\frac{\partial F}{\partial y} = -y\frac{1}{y} - \ln(y) + \lambda = -1 - \ln(y) + \lambda.$$

Since F has no y' dependence, then $\partial F/\partial y' = 0$ and the Euler-Lagrange equation becomes

$$-1 - \ln(y) + \lambda = 0,$$

or

$$\ln(y) = \lambda - 1,$$

which says

$$y = e^{\lambda-1}, \qquad \text{a constant (because } \lambda \text{ is a constant).}$$

We can calculate the value of this constant from the integral constraint:

$$\int_0^M y(x)\, dx = 1 = \int_0^M e^{\lambda-1}\, dx = e^{\lambda-1}\, M.$$

Thus,

$$y(x) = e^{\lambda-1} = \frac{1}{M}, \qquad 0 \le x \le M$$

$$= 0, \qquad\qquad\qquad \text{otherwise.}$$

Now, how do we know this $y(x)$ gives a maximum J, and not a minimum? Because it is easy to demonstrate a different $y(x)$ that gives a J *smaller* than the above solution $y(x)$ gives, i.e., the solution $y(x)$ does *not minimize J*. For the solution $y(x)$, we have

(continued)

$$J = -\int_0^M \frac{1}{M} \ln\left(\frac{1}{M}\right) dx = -\frac{1}{M} \ln\left(\frac{1}{M}\right) M = -\ln\left(\frac{1}{M}\right) = \ln(M).$$

Suppose, however, that we let

$$y(x) = \frac{2}{M}, \qquad 0 \le x \le \frac{1}{2} M$$
$$= 0, \qquad \text{otherwise,}$$

which clearly satisfies the given constraints. For this $y(x)$, we have

$$J = -\int_0^{\frac{1}{2}M} \frac{2}{M} \ln\left(\frac{2}{M}\right) dx = -\frac{2}{M} \ln\left(\frac{2}{M}\right) \frac{1}{2} M = -\ln\left(\frac{2}{M}\right)$$
$$= \ln(M) - \ln(2) < \ln(M).$$

Historical update: The calculus of variations applications that I have discussed in this chapter are the ones of historical interest, but its use today has gone far beyond that of studying beads sliding down wires, and making fences of fixed length to enclose the maximum land. Today's applications have mathematical structures far more complex than I have treated here, with differential equations and inequalities serving as the constraint conditions. Such problems routinely occur in what is called *optimal control theory*. To read much of the modern literature in that subject requires much more background than I have assumed here, but interesting exceptions can be found.

For example, consider the question of how a human runner should vary (i.e., *control*) her speed $v(t)$ during a race of given distance D in order to minimize her running time T. The runner starts from rest, of course, and so $v(0) = 0$. This problem was beautifully analyzed by Joseph B. Keller in "Optimal Velocity in a Race"

(*American Mathematical Monthly*, May 1974, pp. 474–80). Keller begins by writing the mathematical statement

$$D = \int_0^T v(t)\, dt,$$

along with Newton's second law of motion ("$F = ma$") as

$$\frac{dv}{dt} + \frac{v}{\tau} = f(t),$$

where v/τ is the resistive force per unit mass of the runner (τ, called a *physiological constant*, is characteristic of the particular runner) and $f(t)$ is the propulsive force per unit mass of the runner. The runner controls $f(t)$ (and, hence, $v(t)$) with the constraint that there is some maximum force, F, that she can exert (F is another physiological constant characteristic of the particular runner).

Keller next writes $E(t)$ as the oxygen available (per unit mass) to the runner's muscles, and argues that oxygen is *used* at a rate proportional to the product fv (more oxygen is used the faster she runs and/or the harder she tries to run). On the other hand, oxygen is *supplied* at a rate proportional to yet another physiological constant, σ, which measures the efficiency of the lungs and of the blood circulation of the particular runner. That is,

$$\frac{dE}{dt} = \sigma - fv.$$

Finally, the runner is modeled as starting with an initial oxygen level E_0, where of course we demand that $E(t) \geq 0$ (think of what an $E(t) < 0$ would mean, physically, for the runner!).

So, Keller's problem reduces to the following mathematical question:

given the positive constants $\tau, F, \sigma, E_0,$ and D, find $v(t)$ and $f(t)$ such that the T in

$$D = \int_0^T v(t)\, dt, \qquad v(0) = 0$$

is minimized, subject to the constraints that

1. $(dv/dt) + (v/\tau) = f(t)$;
2. $f(t) \leq F$ (Keller doesn't say $f(t) \geq 0$, but I consider that to be an obvious requirement);
3. $dE/dt = \sigma - fv$, where $E(0) = E_0$, $E(t) \geq 0$.

For the (lengthy but nicely explained) details for how to solve this fascinating calculus of variations problem, and how to use the solution to "explain" some world records in long-distance track-and-field, see Keller's paper.

7.

The Modern Age Begins

7.1 The Fermat/Steiner Problem

With the development of the calculus of variations well under way as mathematics entered the nineteenth, attention was redirected to an old problem in geometry. The problem is deceptively simple, but it proved to be a signpost to the future for extrema studies: given a triangle, as shown in figure 7.1, where is the point P inside that triangle that minimizes the sum of the distances from P to the three vertices? P is often called *Steiner's point*, after the nineteenth-century Swiss mathematician Jacob Steiner, whose geometric work on the isoperimetric problem was discussed in section 2.3. In fact, however, the question about P greatly predates Steiner. Indeed, it was originally posed two centuries before Steiner, by Fermat, in his 1629 *Method for Determining Maxima and Minima and Tangents to Curved Lines*. Torricelli in Italy (see the preface again) read this, and took up the challenge.

We know that sometime before 1640 Torricelli was successful in locating P—and so it is occasionally called *Torricelli's point* (often called *Fermat's point*, too)—because his student Vincenzo Viviani (1622–1703) published his late mentor's geometric solution in his book *De maximis et minimis* (1659). It was their fellow Italian, Bonaventura Francesco Cavalieri (1598–1647), however, who is given credit for publishing in 1647 the following interesting property of P, a property that reminds us immediately of one of Plateau's rules for soap films: the lines connecting P to the three vertices meet at P at 120° angles, as long as all three of the vertex angles are each less than 120°. This result is often so very useful in modern

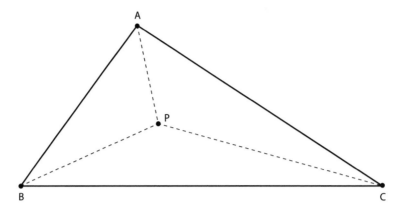

FIGURE 7.1. Steiner's problem.

applications called *facility location problems* (e.g., where to locate the town fire department), that a derivation of the result is instructive. Curiously, a number of published analyses of P use the power of calculus to derive the 120° property, but fail to show where P actually *is*. This is odd because there is a beautiful but elementary geometric proof (different from Torricelli's) that both deduces the 120° rule as well as shows how to locate P. What follows is based on that elegant proof, due to German historian of mathematics J. E. Hoffmann (1900–73), who published it in 1929.

   With reference to figure 7.1, rotate the triangle APB counterclockwise around B by 60°, to arrive at $C'P'B$, as shown in figure 7.2. P rotates into P' and, in particular, PA rotates into $C'P'$, AB rotates into BC', and PB rotates into $P'B$. Thus,

$$PA + PB + PC = C'P' + PB + PC.$$

   Since $PB = P'B$ then the triangle PBP' is isosceles, which means the base angles $\angle BP'P$ and $\angle BPP'$ are equal. But since the third angle of the triangle PBP' is 60° (*by construction*), then all three angles of the triangle PBP' are equal (to 60°), and so triangle PBP' is more than just isosceles—it is equilateral. (By the same sort of argument, so is triangle $AC'B$.) Thus $PB = P'P$. So,

$$PA + PB + PC = C'P' + P'P + PC.$$

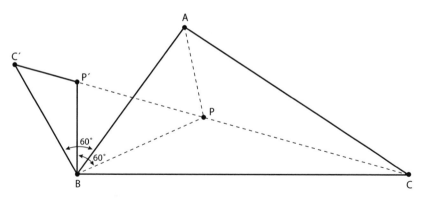

FIGURE 7.2. The 120° rule.

The right-hand side of this equality is, in general, a broken path from C' to C, which of course is *shortest* when it is *straight*. That would require $\angle BPC + \angle BPP' = 180°$, or

$$\angle BPC = 180° - \angle BPP' = 180° - 60° = 120°.$$

Since we could equally well have drawn AC or AB as the horizontal side of the triangle in figure 7.1, we can conclude, too, that

$$\angle APC = \angle APB = 120°.$$

The beauty of Hofmann's analysis is that, in addition to deducing the 120° property, it also shows us how to actually locate P. Here's how.

As stated before, the triangle $AC'B$ is equilateral, and it is easily constructed. If we now construct the analogous equilateral triangle on either of the other two sides of the original triangle (as shown in figure 7.3)—remember, any one of the three sides of triangle ABC could be the one drawn horizontally—then not only will the straight line joining C' to C pass through P but so will the straight line joining the outermost vertex of the second constructed equilateral triangle and the opposite vertex of the triangle ABC. Thus, the intersection point of those two lines *is P*! These two lines (as well as the third line connecting the third equilateral triangle's outermost vertex to its opposite vertex of the triangle ABC) all three through P.

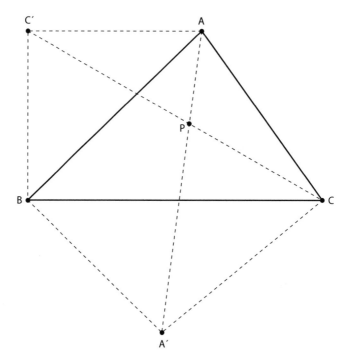

FIGURE 7.3. Locating P when P is inside the triangle.

And finally, this elegant analysis also does something else for us—it tells us that, if one of the angles of the original triangle ABC is equal to or greater than 120°, then the point P is *not inside* the triangle. It is easy to see this because, as the vertex angle at B increases toward 120°, the point P moves toward B and, when the vertex angle reaches 120° the point P *is* B. But what if the angle at B increases beyond 120°? What happens then to P? It is not hard to show in that case that P *remains* at B. Here's why.

Just to be different from the previous discussion, which was based on Hofmann's proof, let's now assume it is the angle at vertex A that is greater than 120°, and that the point P is *outside* the triangle ABC (as shown in figure 7.4). By the first assumption, the angle β between the side AB and the straight-line extension of the side AC is less than 60°. Now, rotate the triangle APB clockwise through the angle β. Thus P rotates into P' and B rotates into B', where it is clear that B' is on the straight-line extension of the side AC. Also,

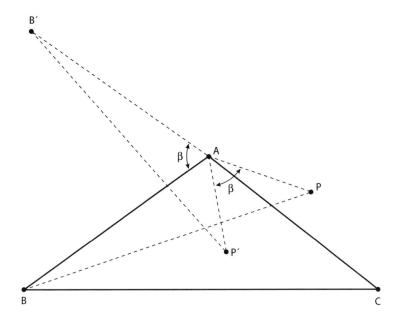

FIGURE 7.4. When P is *not* inside the triangle.

it is equally obvious that $PB = P'B'$ and that $PA = P'A$. Thus, the triangle PAP' is an isosceles triangle with angle β at vertex A. Since $\beta < 60°$, then the equal base angles in that isosceles triangle are each greater than 60°, which says the base side $PP' < PA$. Thus, the quantity we are trying to minimize, $PA + PB + PC$, satisfies the inequality

$$PA + PB + PC > PP' + P'B' + PC,$$

where the right-hand side is the length of the broken-line path $B'P'PC$, which in turn is at least as long as the straight-line path $B'A + AC$. That is,

$$PA + PB + PC \geq B'A + AC.$$

Equality is achieved, i.e., the sum $PA + PB + PC$ is minimized, when $P = A$, and we are done.

The Fermat/Steiner problem is important today because it (or generalizations of it) appear in many interesting problems of our technological society. One such generalized version is called the *factory problem*, for example. To make a direct connection to the Fermat/Steiner problem, suppose there are three factories labeled A, B, and C whose locations mark the vertices of a triangle. These factories are to be supplied with monthly shipments of a crucial part from a soon-to-be-built central warehouse. If the cost of shipping a load of parts from the warehouse to any distant point is directly proportional to both the number of parts shipped and to the shipping distance, then where should the warehouse be located? If the warehouse is at P, and if factory A, factory B, and factory C need a, b, and c parts per month, respectively, then we obviously wish to locate P to minimize the quantity $a(PA) + b(PB) + c(PC)$. If $a = b = c$ then we have the original Fermat/Steiner problem, but if this condition does not hold then we have a more difficult problem. That is, suppose we increase the number of factories from three to n, and write the parts required each month, by each factory, as $c_1, c_2, \cdots, c_n$. Now we have to locate P by minimizing the quantity $\sum_{i=1}^{n} c_i (PX_i)$, where PX_i is the distance between P and factory X_i. For a clever (but lengthy) outline of the general solution for the $n = 3$ case (c_1, c_2, and c_3 are not necessarily equal), see Irwin Greenberg et al., "The Three Factory Problem" (*Mathematics Magazine*, March-April 1965, pp. 67–72).

As a final (and amusing) example of the Fermat/Steiner problem, consider the following little tale. Some years ago (1960), while constructing private telephone networks to connect multiple locations operated by a single business customer, the Bell Telephone Company had to deal with a curious government regulation on how much Bell could charge for the use of a network. Rather than basing its charges on the actual usage of the network (calls per month), they were to be calculated in proportion to the length of the *minimum* length of wire needed to construct a network that could link all the distinct customer locations, even if that wasn't the way the network was actually constructed. One of Bell's customers was Delta Airlines, which had three airport sites (Atlanta, Chicago, and New York City) to be linked. Those sites just happen to form (approximately) the vertices of an equilateral triangle, as shown in figure 7.5, and Bell concluded that its charges should thus be based on a path of wire with length 2 (measured in arbitrary units), shown in the solid line.

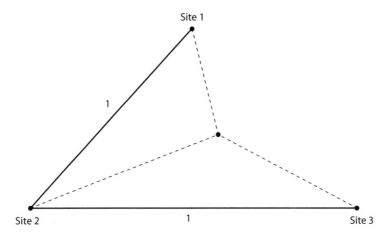

FIGURE 7.5. Geometry of the Delta Airlines problem.

Delta complained, arguing that would result in an overcharge. Somebody at Delta had remembered the Fermat/Steiner problem! What if, Delta asserted, it simply opened a fourth site, a *ghost hub*, at the Fermat point of the three-airport triangle? Then a network linking the three real sites and the ghost site could be built, as shown in the dashed lines of figure 7.5. That network, which we have just seen is of minimum length, has length $\sqrt{3} = 1.732$ (Delta's hypothetical path is called the *Steiner span* of the equilateral triangle). That is, Delta claimed (correctly) that the minimum-length network that *could* be built was 13.4% shorter than Bell's proposed network, and that its billing charges should be correspondingly reduced (Delta was being overcharged by 15.5%). Bell Lab mathematicians, of course, knew a good argument when they saw it and agreed.

If Delta had wanted to link *four* airport sites that had just happened to lie on the vertices of a square, then we can see that the savings achieved by the Steiner span are less dramatic but still significant. Extending Bell's original network idea to four points would give a path length of 3 (in arbitrary units), as shown in the solid line of figure 7.6. The Steiner span, however, in the dashed lines, uses two ghost hubs to achieve a length of $1 + \sqrt{3} = 2.732$, a reduction of 8.9%.

Other problems, similar in spirit to the Fermat/Steiner problem, are treated in the paper by Bennett Eisenberg and Samir Khabbaz,

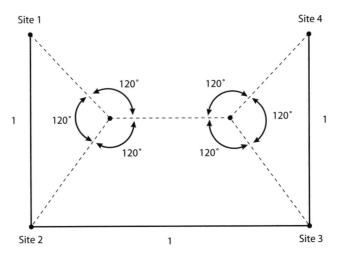

Figure 7.6. An extension of the Delta Airlines problem to two ghost hubs.

"Optimal Locations" (*The College Mathematics Journal*, September 1992, pp. 282–89). One that is particularly interesting concerns the optimal location of the transmitting antenna for a radio station serving several towns. If we make the reasonable assumption that the received signal strength *decreases* with *increasing* distance from the antenna, and further, that we demand the signal strength at the town most distant from the antenna be as strong as possible, then we have the following mathematical problem. If A represents the location of the antenna, and D_i is the distance between A and town i, then we want to position A so that the *maximum* of the D_i is *minimized*.

7.2 Digging the Optimal Trench, Paving the Shortest Mail Route, and Least-Cost Paths through Directed Graphs

For a path-length-minimization problem of an entirely different nature than that of the Steiner/Fermat problem, consider the following passage that I have taken from a 1986 paper (citation to follow soon):

A telephone company, while repairing buried cable, has discovered that although the cable is buried 1 m deep, often the cable is not directly under the marker that is supposed to be erected above it. They do know that the cable is always within 2 m of the marker in the horizontal plane. To ensure finding the cable, even when its direction is unknown, the repairmen dig a 1-m-deep trench in a circle of radius 2 m about the marker.

The geometry of this situation is shown in figure 7.7.

In 1974 it was speculated that a trench with length shorter than the circumference of a circle (but still ensuring the discovery of the cable) could be dug as shown in the solid line of figure 7.7. It has length $2\pi + 4$, and so the ratio of that length to the circumference of the circular trench is

$$\frac{2\pi + 4}{4\pi} = \frac{\pi + 2}{2\pi} = 0.81831,$$

i.e., the shorter trench is more than 18% shorter. It wasn't until 1984, however, that this shorter trench geometry was *proven* to be the shortest possible trench that is a continuous arc. Amazingly,

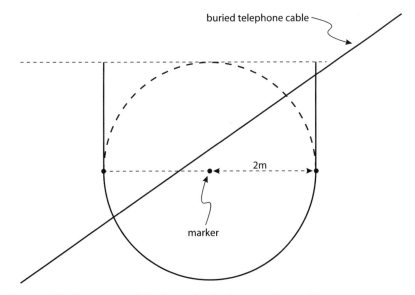

FIGURE 7.7. Geometry of the buried telephone cable problem.

if one allows the trench to be dug as several discontinuous (i.e., unconnected) parts, then this ratio can be further reduced to 0.7639. For the details elaborating on all of these statements, see the paper by V. Faber and J. Mycielski, "The Shortest Curve That Meets All the Lines That Meet a Convex Body" (*American Mathematical Monthly*, December 1986, pp. 796–801).

Another minimal-length problem, in a different context, is that of finding a shortest *closed* path such as is illustrated in figure 7.8. The path is to start on the west side (W) of a quadrilateral (the solid line) at the given point p, and is to eventually return to p (the points A, B, C, and D, the vertices of the quadrilateral, are also specified). The path is constrained only by the requirement that it connects to each of the other three sides of the quadrilateral. *Where* the path actually touches the S, E, and N sides is unspecified—only that the total path length be minimum.

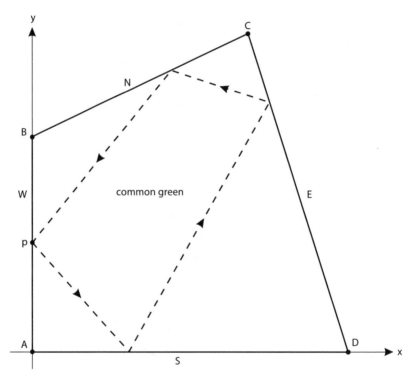

FIGURE 7.8. Shortest interior path around a quadrilateral that visits each side.

One might imagine, for example, that the N, E, and S sides (BC, CD, and AD) of the quadrilateral represent the front property lines of three proposed new homes to be built facing onto a common green, and that p will be on the street entrance (AB) to the common green. The common green is, by covenant, to contain nothing but grass and a closed-path brick walkway allowing access to each of the three homes. Before laying out the walkway, the builder receives a request, from the post office, to make the walkway of minimum length, thus allowing the mail carrier to make his daily journey around the common green in minimum time. (Another example of the Postal Service ever striving for maximum efficiency!) The builder likes this request, too, since it minimizes the number of bricks he has to lay.

There is a very clever solution to this problem, given in the paper by R. A. Jacobson and K. L. Yocom, "Shortest Paths within Polygons" (*Mathematics Magazine*, November-December 1966, pp. 290–93). If we call the quadrilateral Q, and if we then reflect Q through side S to get quadrilateral Q_1 (see figure 7.9), and then continue reflecting

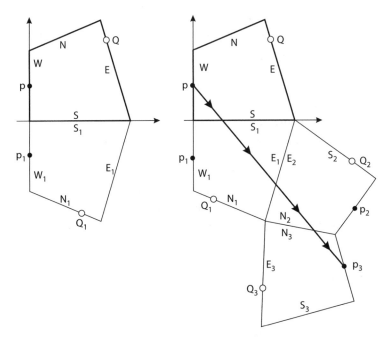

FIGURE 7.9. Finding the shortest path by reflection.

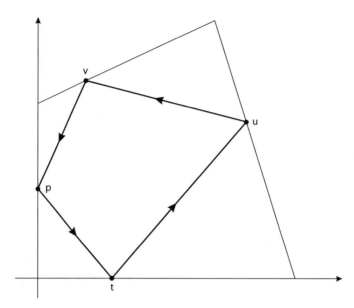

FIGURE 7.10. The shortest path is a straight line.

(i.e., Q_1 is reflected through E_1 to get Q_2, and Q_2 is reflected through N_2 to get Q_3), we then can follow point p through the reflections (marked as p_1, p_2, and p_3). The solution is now obvious—the *shortest* path connecting p to p_3 (a closed loop) is the *straight* path. Where this path crosses $S(= S_1)$, $E_1(= E_2)$, and $N_2(= N_3)$ determines the touching points t, u, and v, respectively (see figure 7.10).

As another minimum-path-length problem of yet an entirely different form, consider the *directed graph* of figure 7.11. That graph has $n = 8$ *nodes* (the circled numbers) that are connected with arcs that are always directed from left-to-right, i.e., from a lower-numbered node to a higher-numbered node. Associated with each arc is a nonnegative number, called the *cost* of that arc, i.e., the cost of traveling from the lower-numbered node to the higher-numbered node. This cost may be the actual distance between the two nodes (in some arbitrary units), or perhaps it is a measure of the *difficulty* of traversing the arc (as measured by some means of the analyst's choosing). We will impose only one constraint on a directed graph: the nodes *must* be numbered in such a way that if node i and node j are linked by an arc *from i to j*, then $i < j$. This restriction prevents the existence of endless closed-loop subpaths within the directed graph.

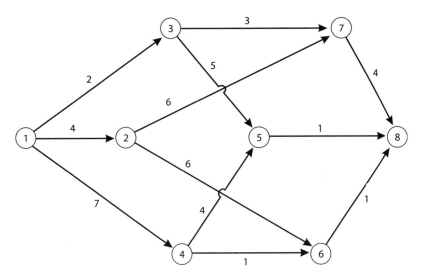

FIGURE 7.11. A directed graph.

The problem is to determine, among all of the possible ways to travel from node 1 at the far left to node 8 at the far right (more generally from node 1 to node n), which path has the minimum total cost. The total cost of a path is defined to be the sum of the costs of the individual arcs that form the path. For example, the total cost of the path $1 \to 3 \to 7 \to 8$ is 9. But is that the minimum-total-cost path? The answer is *no*—can you see which path *is* the minimum-total-cost path? Even if you can, what if instead of $n = 8$, we had a directed graph with $n = 100$ nodes (or 10,000 nodes)? Such large directed graphs would clearly present enormous computational challenges if you resorted to a brute-force enumeration and comparison. I'll not solve this problem now, but instead let you think about it for a while. In section 7.5, as an illustration of dynamic programming, I'll show you how to easily find the minimum-total-cost path, *by hand*, even for pretty large values of n. With the aid of a computer, directed graphs with very large values of n are just as easily processed.

We can see why something better than a simple comparison of the total costs of all possible paths is required by calculating what computer scientists call the *computational burden* of enumeration. To make this calculation general, and not specific for any particular directed graph, let's imagine that every node i links to *every* node

$j > i$ (with the exception, of course, that node n is the end of the path). We'll call such a directed graph a *complete* directed graph. (In section 7.5 I'll show you how directed graphs occur, in a natural way, in a modern production control problem.) In any particular directed graph where a linking arc doesn't actually exist between two nodes, we can effectively remove that arc by simply giving it an extremely large cost, which means the minimum-total-cost path will surely not include that particular arc. The number of arcs, N, in a complete directed graph is easy to calculate. Since node 1 connects to all of the remaining $n - 1$ nodes, and since node 2 connects to all of the remaining $n - 2$ nodes, etc., we have

$$N = (n - 1) + (n - 2) + \cdots + 1,$$

a sum well known to be $\frac{1}{2} n(n - 1)$. (Gauss is said to have done this calculation at age ten!) More subtle, however, is the calculation of the number of paths through those arcs, from node 1 to node n, in a complete directed graph.

To calculate this, let's define $f(i)$ to be the number of paths from node i to node n (the answer to our question is $f(1)$). It is clear, to start, that since at node $n - 1$ we can go only to node n, then

$$f(n - 1) = 1.$$

What is $f(n - 2)$? Well, at node $n - 2$, we can go to just two places; directly to node n or to node $n - 1$. Thus,

$$f(n - 2) = 1 + f(n - 1) = 1 + 1 = 2.$$

What is $f(n - 3)$? Well, at node $n - 3$, we can go to just three places; directly to node n or to node $n-1$ or to node $n - 2$. Thus,

$$f(n - 3) = 1 + f(n - 1) + f(n - 2) = 1 + 1 + 2 = 4.$$

One more time. What is $f(n - 4)$? Well, at node $n - 4$, we can go to just four places; directly to node n or to node $n - 1$ or to node $n - 2$ or to node $n - 3$. Thus,

$$f(n - 4) = 1 + f(n - 1) + f(n - 2) + f(n - 3) = 1 + 1 + 2 + 4 = 8.$$

The pattern is clear:

$$f(n - k) = 2^{k-1}$$

The answer to our problem, $f(1)$, means $k = n - 1$. So, in a complete directed graph with n nodes, there are a total of

$$f(1) = 2^{n-2} = \frac{1}{4} \cdot 2^n \text{ paths.}$$

The number of paths grows *exponentially* with the number of nodes, and so gets very big, very fast. For example, with just $n = 35$ nodes, there are a total of

$$f(1) = \frac{1}{4} \cdot 2^{35} = 8,589,934,592 \text{ paths.}$$

Would *you* want to compute the total cost of each one of them to find the path with minimum total cost? I didn't think so!

7.3   The Traveling Salesman Problem

A characteristic of a number of modern optimal-path problems is that the existence of a solution is theoretically obvious by inspection, and yet they remain unsolvable in practice. This is in dramatic contrast to the historically important problems discussed in the previous chapters. For those problems it was not at all obvious, by *any* means, what the solutions might be or even if there was a solution. The most famous of the modern optimal-path problems is the so-called "traveling salesman problem," which gets its name from the amusing context in which the problem is usually presented. Imagine that a salesman, who lives in City 0, periodically drives to n other cities to visit clients. He travels to one city after another, visiting each city exactly once, and then after seeing the nth client returns home to City 0. He knows the distance between each pair of cities, and from this knowledge he wishes to determine the particular sequence of city visits that minimizes the total, closed-loop travel distance.

If we write $d(i, j)$ as the distance to travel from City i to City j, then determining the total travel distance for a given ordering of cities is a trivial exercise in addition. (By the way, notice that in general $d(i, j) \neq d(j, i)$, because the roads connecting two cities may be one-way roads of different length. Indeed, if two cities are linked in just one direction we could have the case $d(i, j) < \infty$ and $d(j, i) = \infty$.)

The "solution" to the traveling salesman problem is now obvious —simply look at all permutations of the integers 1 to n that, additionally, *start and end* with 0 (each such string of numbers is a possible closed-loop path that represents a legitimate travel route), compute the total travel distance for each string, and select the string with the smallest total. This approach is certain to find the solution, but the problem with it is that the number of permutations grows at an enormous rate with n. This is because, starting at City 0, the salesman has n choices for the city to visit first, then $n - 1$ choices for the second city to visit, $n - 2$ choices for the third city to visit, etc. Finally, after visiting the last city on his list (after his nth choice), he returns home to City 0. So, there are a total of $n!$ closed-loop candidate paths for a tour of $n + 1$ cities (City 0, plus Cities 1 through n).

For $n = 6$ (7 cities) there are just $6! = 720$ closed-loop paths to compare. But for $n = 70$ (71 cities) there are $70!$ routes to compare, i.e., almost 1.2×10^{100} routes. Increasing the tour by a factor of 10 (7 cities to 71 cities) has resulted in a supernova explosion of the number of tours that must be compared. $n!$ grows far more rapidly than exponential growth, and for $n = 70$, the solution, by brute-force enumeration, has become computationally beyond any computer we can imagine being built using today's technology of clocked, sequential logic. So, while the traveling salesman problem clearly has a solution for any value of n, once n exceeds just a modest value, nobody can actually determine, by enumeration, what that solution actually is!

A dramatic illustration of just how absurd the brute-force enumeration "solution" is was given some years ago by George Dantzig (1914–), the American mathematician who in 1947 developed the astonishingly effective simplex algorithm for solving linear programming problems (the topic of the next section). In his paper "Reminiscences about the Origins of Linear Programming" (*Operations Research Letters*, April 1982, pp. 43–48), Dantzig wrote

Now 70! is a big number, greater than 10^{100}. Suppose we had an IBM 370-168 [a very big computer in the 1980s] available at the time of the Big Bang 15 billion years ago. Would it have been able to look at all the 70! combinations by the year 1981? No! Suppose instead it could examine 1 billion assignments per second? The answer is still no. Even if the Earth were filled with such computers all working in parallel, the answer would still be no. If, however, there were 10^{50} earths or 10^{44} suns all filled with nanosecond speed computers all programmed in parallel from the time of the Big Bang until the Sun grows cold, then perhaps the answer is *yes*.

A large number of important problems that commonly occur in modern society have this same property of enormous computational complexity if attacked head-on by brute-force enumeration. For example, in the above illustration, Dantzig was writing not of the traveling salesman problem but rather of the so-called "assignment problem": how to assign 70 men to 70 jobs (with each man providing different skill levels for each job) in such a way as to get all 70 jobs done in minimum time. Because of the "$n!$ problem" much effort has gone into discovering *computationally efficient* algorithms. Two general approaches that have been developed are the one I just mentioned, called *linear programming* (Dantzig's simplex algorithm could, in 1981, solve the 70! assignment problem in less than a second), and the very different method of *dynamic programming*. To finish this book I'll briefly discuss each in the next two sections.

7.4 Minimizing with Inequalities (Linear Programming)

In this section you'll encounter the fundamental ideas of linear programming, a topic on which literally hundreds of books have been written over the last fifty years, from very elementary treatments to ones using mathematical techniques far beyond the level of this book. The presentation, here, in a single section, will necessarily be at the most basic level, along with some historical commentary. But let me clear away one common misunderstanding, immediately, before I begin. Linear programming is a mathematical technique that can be (usually is) implemented as a computer program, but that is not what the word "programming" means. Rather, its original historical usage was as the name for the administrative task of scheduling a sequence of time-ordered events (usually with the objective of optimizing some measure, e.g., minimizing the total cost, or the total time required). Indeed, the naming of the task of writing code for a computer as *programming* derives from that historical origin, as after all that is what writing a computer program *is*—the scheduling of a time-ordered sequence of events, with each event being the execution of an individual instruction in the program code. Programming (i.e., scheduling) problems, however, were studied long before the first programmable electronic computers were built.

The formal definition of the mathematical linear programming problem is quite simple, in principle:

> given a linear function f (called the *objective function*) of n independent, nonnegative real variables $x_1, x_2, \cdots, x_n$, along with a system of inequalities linear in the x_i, find the specific values of the x_i that minimize (maximize) f.

A vast number of important optimization problems in modern society have this structure. The mathematical study of systems of inequalities can be traced as far back as 1826, to the French mathematician Joseph Fourier (1768–1830). That year he published a short paper in which he considered a problem involving multiple inequalities that, together, define what he called an "irregular polygon" and what is today called a convex region. He elaborated on these ideas in a second paper published the following year. Fourier's work was a fundamental foreshadowing of concepts basic to modern linear programming. You can find more on what he did in two papers by I. Grattan-Guiness, "Joseph Fourier's Anticipation of Linear Programming" (*Operational Research Quarterly*, 1970, pp. 361–64), and "On the Prehistory of Linear and Non-Linear Programming," in *The History of Modern Mathematics*, vol. 3 (Academic Press 1994, pp. 43–89), and in the paper by H. P. Williams, "Fourier's Method of Linear Programming and Its Dual" (*American Mathematical Monthly*, November 1986, pp. 681–95).

The concepts of linear programming optimization are no longer limited to just scholarly journals, but have actually penetrated into popular fiction as well. For example, in Robert K. Tanenbaum's 1987 novel *No Lesser Plea*, we find the following little speech from a lawyer in the San Francisco District Attorney's Office:

> To gain maximum efficiency, we have to view the entire criminal justice system as a whole, and adjust the inputs of resources at each node so as to optimize throughput. . . . so we have developed a Trial Screening Profile that assigns priorities to different sorts of cases and generates scores. Then we can observe the trial dispositions of various [assistant district attorneys] and bureaus and see whose scores diverge from the optimum and take corrective action.

Later in the same novel we get a skeptical response from another lawyer who heard the above:

> Look, they're trying to control the whole office with numbers. But you can't really control anything with numbers unless you have a sense of what the numbers mean. Which they don't. . . . It's like that story about the Russian chandelier factory. They get a quota from Moscow each year—make six tons of chandeliers. So they make one six-ton chandelier and take the rest of the year off.

As the first example of a linear programming problem, the famous "diet problem" formulated by the American economist George Stigler (1911–91) is, I think, the best choice. The diet problem is a mainstay in today's textbook discussions of linear programming, both because it is obviously important and easy to understand. It is *not* generally easy to solve, however, at least not by hand. Imagine that we have two lists in front of us. One gives a number of different foods, their nutritional content (vitamins, minerals, fiber, fat, calories, etc.) per unit amount, as well as the cost of each food per unit amount. The other list contains the minimum and/or maximum amounts of nutritional intake, per unit time, required by an adult to maintain good health. The solution to the diet problem is the determination of the amount of each food required, per unit time, to satisfy the nutritional needs *at minimum cost*.

In a paper published in 1945 ("The Cost of Subsistence," *Journal of Farm Economics*, pp. 303–14), Stigler attempted to solve the diet problem using actual data for Americans. For a total of 77 available foods in August 1939, along with nine nutritional constraints, he arrived at a diet with a yearly cost of $39.93. It was a pretty awful diet (wheat flour, evaporated milk, cabbage, spinach and beans!), of which Stigler said "No one recommends [this diet] for anyone, let alone everyone." It does sound a little like the gruel given by Dickens to Oliver Twist but, still, if not particularly tasty, it *is* a low-cost, nutritionally sound diet. But was it the *minimum*-cost nutritionally sound diet? Stigler was careful to not make that claim because, as he wrote, his "[analysis] procedure is experimental because there does not appear to be any direct method of finding the minimum of a linear function subject to linear conditions."

Just two years later, however, Dantzig published just such a method, his now famous *simplex* algorithm. Indeed, in the fall of 1947, the Mathematical Tables Project (MTP) at the National Bureau of Standards used the simplex algorithm to solve Stigler's array of nine inequalities in 77 nonnegative variables. [The MTP was a Depression-era Work Projects Administration effort that employed hundreds of out-of-work office clerks. See David Alan Grier, "The Math Tables Project of the Work Projects Administration: the Reluctant Start of the Computing Era" (*IEEE Annals of the History of Computing*, no. 3, 1998, pp. 33–50)]. Using just desk calculators, it required almost 17,000 multiplications and divisions spread over more than 100 man-days of work to do the job; the least expensive, nutritionally sound diet was determined to cost $39.69 per year, using not Stigler's five foods but instead *nine* (wheat flour, corn meal, evaporated milk, peanut butter, lard, beef liver, cabbage, potatoes and spinach—*still* pretty awful!).

Stigler's accomplishment of getting to within 0.6% of the correct solution to such a complex problem was certainly impressive, indeed amazing, but his trial-and-error approach would have no hope of producing similar success in the face of a problem with *tens of thousands* of variables and constraints. And such monster problems are the *typical* linear programming problem today, occurring in such applications as scheduling airline flights and routing telephone calls (both of which attempt to maximize a *flow* through a global network with varying local limits on congestion). Dantzig's simplex algorithm, on the other hand, handles such problems with ease. Programmed on a modern home computer, for example, Stigler's original diet problem is solved in the blink of an eye.

The simplex algorithm requires linear algebra to properly explain it, but we can appreciate what it does with the following simplified diet problem. Imagine that, to be healthy and grow into a fine, fat chicken dinner, a chicken needs to consume a minimum weekly amount of three different nutrients each (called A, B, and C). Let's further assume, to be specific, that the minimal amounts are (in some unit system) 60, 84, and 72, respectively. The local feed store stocks two brands of chicken feed, markedly different from each other. One is cheap because it's low on nutrients per ounce, and the other is expensive because it's high on nutrients per ounce. To be specific, suppose the details are

	A	B	C	Cost
	(nutrients per ounce)			(pennies per ounce)
Feed #1	3	7	3	10
Feed #2	2	2	6	4

From this we can see that if the farmer buys just feed #1, then, to provide a chicken with its minimum weekly nutrients, he must buy the maximum of $\{60/3, 84/7, 72/3\}$ ounces = the maximum of $\{20, 12, 24\}$ ounces = 24 ounces, at a cost of $2.40 per week. On the other hand, if the farmer buys just feed #2, then, to provide a chicken with its minimum weekly nutrients, he must buy the maximum of $(60/2, 84/2, 72/6)$ ounces = the maximum of $(30, 42, 12)$ ounces = 42 ounces, at a cost of $1.68 per week. Obviously, if he is going to buy just one of the feeds then feed #2 is the better (i.e., cheaper) *pure* strategy. But is that the best possible choice? That is, could he feed a chicken a nutritionally adequate diet for less than $1.68 per week if he used a mixed strategy, i.e., if he used a mix of the two feeds? We can answer this question by stating the problem as one in linear programming.

If we denote the weekly amount (in ounces) of feed #1 by x_1 and of feed #2 by x_2, then we can express the nutritional constraints with the following three inequalities:

(a) $3x_1 + 2x_2 \geq 60$ (nutrient A)

(b) $7x_1 + 2x_2 \geq 84$ (nutrient B)

(c) $3x_1 + 6x_2 \geq 72$ (nutrient C).

The objective function to be minimized is the weekly cost:

$$f(x_1, x_2) = 10x_1 + 4x_2.$$

The constraint inequalities can be written as

$$x_2 \geq 30 - \frac{3}{2} x_1$$

$$x_2 \geq 42 - \frac{7}{2} x_1$$

$$x_2 \geq 12 - \frac{1}{2} x_1.$$

The geometric meaning of each inequality is easy to understand: if, for example, we plot the line $x_2 = 30 - \frac{3}{2}x_1$ (the equality version of the first inequality), then the *inequality* is satisfied by any point *on* the line or *above* the line. The three inequalities are simultaneously

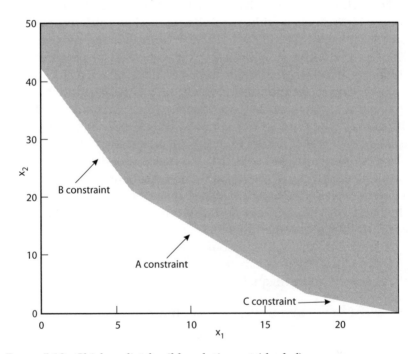

FIGURE 7.12. Chicken diet feasible solution set (shaded).

satisfied by any point that is (the shaded region shown in figure 7.12) above all three lines (or on the boundary edge of that region). All of those points together form the so-called *feasible solution set*. For this problem we see that the feasible solution set is an unbounded convex region in the first quadrant of the x_1, x_2 plane. We also see that the two pure strategies considered earlier are represented by the point (0, 42), which is the pure strategy of using only feed #2, and the point (24, 0), which is the pure strategy of using only feed #1. In addition, there are two other vertex points on the boundary of the convex feasible solution set. One, the point (6, 21), is determined by the intersection of the *A* and *B* lines and the other point, (18, 3), is determined by the intersection of the *A* and *C* lines.

The feasible solution set is often called a *polytope* or *simplex* (hence the algorithm's name), but this is a loose use of technical language. A simplex in *n*-dimensional space is the most elementary (minimum complexity) structure that exists in the *complete* space, e.g., a triangle in two-dimensional space is a simplex, while a line is not because it uses only one of the two dimensions available. An *n*-space simplex has $n + 1$ noncollinear vertices (e.g., a triangle in 2-space has three vertices not all on the same line). A feasible solution set, however, as I'll soon demonstrate, can have *lots* more than $n + 1$ vertices! The concept of the simplex greatly predates linear programming; it was introduced a century earlier as the *prime confine* by the great English mathematician William Kingdon Clifford (1845– 79). Citing the two- and three-dimensional versions as the triangle and the tetrahedron for the "simplest form of confine" of an area and a volume, respectively, Clifford generalized the idea to *n*-dimensions, noting that the prime confine in *n*-space has $n + 1$ vertices.

Let's now plot the objective function on top of the feasible solution set, as shown in figure 7.13, using several different constant values for f. We see that the result is a family of parallel, straight lines with the general equation

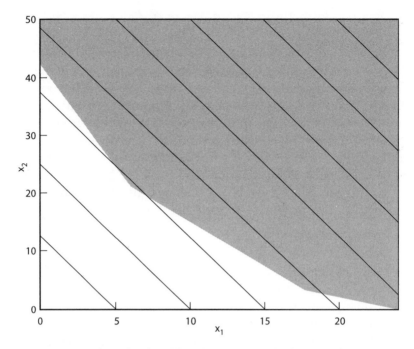

FIGURE 7.13. Chicken diet feasible solution set with objective function.

$$x_2 = \frac{1}{4} f - \frac{5}{2} x_1.$$

The lines are parallel, of course, since each has the same slope of $-\frac{5}{2}$. To solve our problem geometrically, then, we simply look for that objective function line with the smallest value of f that still passes through the feasible solution set. Since the feasible solution set is convex, then the minimum f line will be the line that is as far to the left as possible, i.e., the line that passes through the feasible solution set vertex at $(x_1 = 6, x_2 = 21)$. Thus, the minimum-cost weekly diet for a chicken consists of 6 ounces of feed #1 mixed with 21 ounces of feed #2, at a weekly cost of $(6 \times 10¢) + (21 \times 4¢) = \$1.44$. This is more than 14% cheaper than the cheaper of the two pure-strategy diets.

What happened, geometrically, in this problem is that each constraint inequality divided a two-dimensional space in half with a one-dimensional line. All of those divisions together carved out a convex region (unbounded, in this case) of the two-dimensional space, which we call the feasible solution set. *Any* point in that set satisfies all of the constraints. The minimum of the objective function was located at one of the extreme points of that set, i.e., at a vertex of the feasible solution set. The same thing happens as we encounter problems with more than two variables. Thus, in an n-variable problem, each constraint inequality divides an n-dimensional space in half with an $(n - 1)$-dimensional surface. All of these divisions together carve out a convex region of the n-dimensional space, which is the feasible solution set. The extreme of the objective function is located at one of the extreme, outermost points of that set, i.e., at a vertex of the feasible solution set.

For a two-variable problem like the chicken diet problem, it is easy to literally watch all of this taking place on a flat piece of paper. In n-dimensional space it is not so easy to "see"! And while we do not have to consider all of the points in the feasible solution set, but only those points on the surface that are the vertices of the set, there are nevertheless a *lot* of vertices as n increases into the thousands. In n-dimensional space, the convex feasible solution set resembles the faceted face of an n-dimensional diamond; the simplex algorithm moves over the face from vertex to vertex with the goal of increasing/decreasing the objective function at each move. When that can no longer be done, *the* optimum vertex has been found. *How* the movement from vertex to vertex is controlled *is* the simplex algorithm, and, in general, is explainable only in the mathematics of linear algebra and matrix theory; I refer you to any good book on linear programming—see, for example, the last paragraph of this section.

We can calculate an upper bound on the number of vertices for the feasible solution set as follows: if there are n variables (each defining a nonnegativity constraint of the form $x_i \geq 0$), and m additional constraints, then there may be as many as $\binom{m+n}{n} = (m + n)!/m!\, n!$ vertices. This claim follows from the simple idea of taking any n of the total of $m + n$ constraints and solving them as equalities, thus defining the values of x_i for a *possible* vertex. The reason I say that this is an upper *bound* is because we may find that, for certain selections of the n constraints from the $m + n$ total constraints, a true vertex is not defined. For example, in the chicken diet problem we had $n = 2$ variables and $m = 5$ constraints (the $x_1 \geq 0$ and $x_2 \geq 0$ ones, plus the three nutritional constraints). For a problem of this size, the upper bound on the number of feasible solution set vertices is

$$\frac{7!}{5!\,2!} = 21,$$

but in fact we found in the chicken diet problem that there are just four vertices (look back at figure 7.12). A potential vertex that didn't make the final cut is, for example, the solution of the equation versions of the nutritional constraint inequalities for B and C. Their solution point is clearly *not* a vertex of the feasible solution set. But even if only a very small fraction of the upper bound is realized, the number of vertices can be enormous. For example, in Stigler's original diet problem with $n = 77$ variables and $m = 9$ nutritional constraints, the upper bound is nearly half-a-*trillion*, i.e.,

$$\binom{86}{9} = \frac{86!}{77!\,9!} = 4.6 \times 10^{11}.$$

For a second example of what appears to be linear programming (but actually isn't), imagine a large industrial production facility with (initially) 1,000 identical operational machines, each of which can make either of two parts (I'll call the two parts A and B). The manager of the facility can assign each operational machine, each week, to the task of making either part A or part B. The manager's goal is to maximize the total profit generated by his facility over the next four-week period. Part B generates more profit than does Part A, but mechanically stresses a machine more than does Part A. He has to decide how to assign his operational machines, at the start of each new week, with the following constraints:

1. if a machine makes part A for a week, then that machine will generate a profit for that week of \$400.
2. if a machine makes part B for a week, then that machine will generate a profit for that week of \$600.
3. of all the operational machines assigned each week to make part A, 20% will suffer some mechanical breakdown and be unavailable for future assignment.
4. of all the operational machines assigned each week to make part B, 40% will suffer some mechanical breakdown and be unavailable for future assignment.

To formulate the manager's problem in mathematical terms, let's write

x_i = number of operational machines assigned to make part A during week i, $i = 1, 2, 3, 4$

y_i = number of operational machines assigned to make part B during week i, $i = 1, 2, 3, 4$

and the objective function (the total, four-week profit) as

$$f = 400 \, (x_1 + x_2 + x_3 + x_4) + 600 \, (y_1 + y_2 + y_3 + y_4) \, .$$

The manager's problem is to determine the *integers* x_1, x_2, x_3, x_4, y_1, y_2, y_3, and y_4 that maximize f, subject to the following constraints:

$$x_1 + y_1 = 1000$$

$$x_2 + y_2 = 0.8 \, x_1 + 0.6 \, y_1$$

$$x_3 + y_3 = 0.8 \, x_2 + 0.6 \, y_2$$

$$x_4 + y_4 = 0.8 \, x_3 + 0.6 \, y_3$$

$$x_i \geq 0, \qquad i = 1, 2, 3, 4$$

$$y_i \geq 0, \qquad i = 1, 2, 3, 4.$$

The requirement that the x_i, y_i be integers is, of course, the result of the obvious condition that operational machines come only in integers. This is a requirement that was not present in the diet problem, and it dramatically alters the mathematical structure of the problem. One can attempt to use linear programming to find the x_i, y_i, and *if* the result happens to give them as integers, then all is well. But there is no guarantee that will happen. Indeed, when it does happen, it is simply a lucky accident of the numbers. I'll solve this problem in the next section, with the entirely different method of dynamic programming, which will automatically give us the integer solution.

One might naively hope that, if linear programming arrives at a noninteger solution, then perhaps simply rounding that solution to integers will solve the problem. That, unfortunately, is not generally true, as shown by a simple counterexample. Suppose we wish to maximize the objective function $f = 40\,x_1 + 70\,x_2$, subject to the constraints $3\,x_1 + 5\,x_2 \le 15$ and $x_1 + 5\,x_2 \le 10$, as well as that x_1 and x_2 must be nonnegative integers. If we initially ignore the integer constraint, then we will arrive at the shaded convex region of figure 7.14 as the feasible solution set (for now, ignore the circled points). Plotting a family of parallel objective function lines shows that the one giving the maximum f that still intersects the feasible solution set is the one that passes through the vertex at (2.5, 1.5), i.e., $x_1 = 2.5$ and $x_2 = 1.5$. Thus,

$$f_{\max} = 40(2.5) + 70(1.5) = 100 + 105 = 205.$$

Rounding the linear programming solution to the integer-coordinate points in the feasible solution set (to the points (3,1) and (2,1)) might seem like the next thing to do, but neither of these is the solution to the so-called *integer*-programming problem. Here's why.

For the problem at hand, we can find the integer solution by enumeration, i.e., by simply testing all points in the feasible solution set with integer coordinates. There are 11 such points, the points circled in figure 7.14. For this small-scale two-dimensional problem, enumeration is only a bit tedious (in problems with more variables, you can see matters would get *very* tedious, very quickly):

Integer Coordinates	$f = 40\,x_1 + 70\,x_2$
(0, 0)	0
(0, 1)	70
(0, 2)	140
(1, 0)	40
(1, 1)	110
(2,0)	80
(2,1)	150
(3,0)	120
(3,1)	190
(4,0)	160
(5,0)	200

Thus, the solution to the integer programming problem is $x_1 = 5, x_2 = 0$, which gives $f_{\max} = 200$. This solution is not even close to

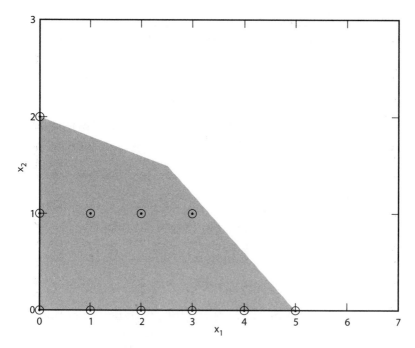

FIGURE 7.14. Integer programming.

the solution to the linear programming problem ($x_1 = 2.5$, $x_2 = 1.5$), or to its possible "rounded" values, and so rounding is discredited.

It wasn't until 1958 that the American mathematician Ralph Gomory (1929–), then an assistant professor of mathematics at Princeton, published an algorithm for solving the integer programming problem. You can find Gomory's breakthrough idea (the so-called *method of cuts*) discussed in the tutorial paper by Joe Wampler and Steve Newman, "Integer Programming" (*The College Mathematics Journal*, March 1996, pp. 95–100). In 1991 Gomory wrote a fascinating historical essay on how he came to integer programming, and you can find that paper ("Early Integer Programming") reprinted in the January–February 2002 issue of *Operations Research*, (pp. 78–81). His original motivation came from a chance remark he heard during a lecture on the Navy's use of linear programming to study its ship assignments within the fleet: getting answers like "1.3 aircraft carriers" were of little value for planners!

Linear Programming and the Nobel Prize in Economics

The economics Nobel prize is not one of the original prizes, but rather is formally "The Bank of Sweden Prize in Economic Sciences in Memory of Alfred Nobel," first awarded in 1969. In 1975 the economics prize was shared by the Soviet mathematician Leonid Kantorovich (1912–86) and the Dutch-born American economist Tjalling Koopmans (1910–86). The citation on their award was "for their contributions to the theory of optimum allocation of resources," i.e., for linear programming. Those two men certainly were of the caliber one would expect for a Nobel laureate, but what of George Dantzig, who is the recognized *inventor* of the simplex algorithm? (Dantzig's 1975 National Medal of Science specifically cites him as the inventor of linear programming.) Dantzig was simply passed over by the Nobel awards committee, an act of stunning omission. Even the winners alluded to this, with both mentioning Dantzig in their Nobel speeches. In fact, when George Stigler won the same prize seven years later, in 1982, he too mentioned Dantzig. (Stigler's prize was *not* specifically for his diet problem work, but rather for unrelated analyses in market behavior and public regulation theory.)

Some Nobel observers feel the Nobel awards committee ignored Dantzig because he is a mathematician, while Kantorovich and Koopmans made their marks as economists. That argument is somewhat supported by a statement made by the economist Robert Dorfman in his paper "The Discovery of Linear Programming" (*Annals of the History of Computing*, July 1984, pp. 283–95): "Linear programming is not a branch of mathematics. It lies in the domains of economics (both applied and theoretical) and management." Dorfman's paper is, in many respects, an admirable piece of historical writing; in particular, he nicely explains the motivations behind the work of Kantorovich and Koopmans. But his erroneous "not mathematics" claim did not pass unnoticed. Vigorous letters of rebuttal from pioneers in the use of linear programming were received and printed in the *Annals* (see vol. 11, no. 2, 1989, pp. 145–51). And even before Dorfman's paper appeared, a

counterexample to his assertion had appeared in a *physics* journal, showing how to apply linear programming to solve certain difficult electrical circuits problems: J. N. Boyd and P. N.
Raychowdhury, "Linear Programming Applied to a Simple Cir-
cuit" (*American Journal of Physics*, May 1980, pp. 376–78). The
authors cite *Dantzig's* work, not Koopmans' or Kantrovich's.

The snub argument is, however, somewhat weakened by
noting that Kantorovich's doctorate was in mathematics, and
that the 1994 economics prize was awarded (in part) to mathematician John Nash (of "A Beautiful Mind" fame) for his purely
mathematical work in game theory. Perhaps, in fact, Nash's
prize was partially motivated by a desire on the part of the Nobel awards committee to show there is no bias by economists
against mathematicians. Still, wouldn't it be more direct for
economists to simply honor the man who invented an algorithm used millions of time each day around the world—
mostly by economists? It should be no surprise to learn that
the *top* prize among mathematicians is *not* the economics Nobel, but rather the Fields Medal (often called the "Nobel prize
of mathematics").

As a fabulously successful algorithm, the simplex algorithm had
long been viewed as the gold standard. It had also long been suspected that it is not perfect in the computationally efficient sense.
What that means is this: define the *size S* of a linear programming
problem to be the sum of the number of variables and constraint
conditions. For example, the chicken diet problem discussed earlier
has a size of $S = 2$ (variables) $+3$ (constraint conditions) $= 5$. Now,
a *computationally efficient* algorithm (for *any* problem) is one that
requires, at most, a solution time that increases as some *polynomial*
function of S. The simplex algorithm has been shown, however, in
its worst case, to require an *exponential* solution time. Such worst-
case problems, I should mention, do not actually seem to occur very
often in "real life," and the simplex algorithm almost always works
quite well on problems with sizes up to $S = 20,000$ or more. This
is because, despite being theoretically exponential, it nearly always
performs (i.e., converges to a solution) in a time approximately proportional to S^1. This (desirable) average behavior caught Dantzig's

early attention. In a long interview given some years ago (*College Mathematics Journal*, September 1986, pp. 293–314), he said of the simplex algorithm:

> Most of the time it solved problems with m equations in $2m$ or $3m$ steps—that was truly amazing. I certainly did not anticipate that it would turn out to be so terrific. I had no experience at the time with problems in higher dimensions, and I didn't trust my geometrical intuition. For example, my intuition told me that the procedure would require too many steps wandering from one adjacent vertex to the next. In practice it takes few steps. In brief, one's intuition in higher dimensional space is not worth a damn!

Still, linear programming problems with sizes much larger than 20,000 are becoming increasingly common (tens of thousands of constraints and hundreds of thousands of variables occur in routine "industrial strength" problems), and so the search started decades ago for alternatives to the simplex algorithm, for algorithms that would *always* execute in polynomial time. In 1979, the Soviet mathematician Leonid Khachiyan (now on the computer science faculty at Rutgers University) announced the final step to earlier work (by others) that resulted in what is called the *ellipsoid* algorithm, which always runs in polynomial time (in a time proportional to S^6). But since the simplex algorithm exhibits an S^1 behavior nearly all the time anyway, the 1979 result was mostly of academic interest only. That interest was intense, however, and a wonderfully funny and insightful essay on how even usually serious people went "off the deep end" about the ellipsoid algorithm is by Eugene L. Lawler, "The Great Mathematical Sputnik of 1979" (*Mathematical Intelligencer*, 1980, pp. 191–98).

The nonpolynomial time property of the simplex algorithm wasn't *proven* until 1972, but when it was it was by the most convincing type of mathematical proof there is—the production of specific examples. That year the American mathematicians Victor Klee (1925–) and George Minty (1929–86) published the following *class* of n-variable (pick n to be any integer greater than zero) linear programming problems:

maximize $f = 2^{n-1}x_1 + 2^{n-2}x_2 + \cdots + 2x_{n-1} + x_n$ subject to the n constraints

$$x_1 \leq 5$$

$$4x_1 + x_2 \leq 25$$

$$8x_1 + 4x_2 + x_3 \leq 125$$

$$\vdots \qquad \vdots$$

$$2^n x_1 + 2^{n-1} x_2 + \cdots + 4x_{n-1} + x_n \leq 5^n.$$

There are 2^n vertices on the resulting feasible solution set and, starting at $x_1 = x_2 = \cdots = x_n = 0$ (which obviously both satisfies the
constraints and is a vertex), Klee and Minty showed the simplex algorithm would find the vertex that maximizes f—$(0, 0, \cdots, 5^n)$—as
the *last* vertex. As a simple example of how polynomial and exponential times compare, consider the following table of 2^S and S^6:

S	2^S	S^6
2	4	64
5	32	15,625
10	1,024	1,000,000
20	1,048,576	64,000,000
29	536,870,912	594,823,321
30	1,073,741,824	729,000,000

The exponential time algorithm is actually *faster* ($2^S < S^6$) than the
polynomial time algorithm for $S \leq 29$. Of course, for $S \geq 30$, we
would obviously prefer the polynomial time algorithm. $S = 30$ is,
in fact, actually a pretty *small* linear programming problem.

In 1984, the Indian analyst Narendra Karmarkar (then at AT&T
Bell Laboratories and not yet 30 years old) announced an entirely
different polynomial time algorithm, one that runs in a time proportional to $S^{3.5}$. (It is interesting to note that he was not trained as
a mathematician, but rather as an electrical engineer.) His algorithm
seems to possess the simplex algorithm's property of nearly always
performing better than its absolute worst-case limit when applied
to problems of everyday structure. For problems with sizes much
larger than 20,000, it appears to run from 50 to 100 times faster than
does the simplex algorithm. The Karmarkar algorithm is called an
interior algorithm because, unlike the simplex algorithm, the search
for the extreme vertex starts from *inside* the feasible solution set.

The simplex algorithm, by contrast, remains entirely on the hyper-dimensional surface of that set as it moves from vertex to vertex.

Karmarkar's algorithm was initially viewed with great skepticism, not only because of the dramatic speed claims for its performance, but also because AT&T refused to reveal important details until after it received a patent on the algorithm (in 1988). There was precedent for this—the first U.S. software patent had been granted years before, in 1968. Many computer science observers, however, don't believe patenting software is the best way for computer science research to develop. Dantzig openly published his simplex algorithm, and from that public accessibility came enormous productive research and useful knowledge. But, of course, we see today a parallel legal path being taken in the biotechnology fields, with companies attempting to patent the DNA codes of everything from microbes to humans, the very "algorithms of life"! Today, the detailed theory behind the Karmarkar algorithm is readily available: you can find it, the ellipsoid algorithm, *and* the simplex algorithm, all discussed in solid mathematics in the single book by the mathematician Howard Karloff, *Linear Programming* (Birkhäuser 1991). A very nice discussion of Karmarkar's algorithm, with some interesting applications of it, had already appeared a few years before in the journal literature; see Gilbert Strang, "Karmarkar's Algorithm and Its Place in Applied Mathematics" (*The Mathematical Intelligencer*, vol. 9, no. 2, 1987, pp. 4–10).

7.5 Minimizing by Working Backwards (Dynamic Programming)

In this final section of the book I'll discuss an important mathematical development that occurred almost at the same time as did the start of linear programming. When a problem can be formulated as a time-ordered sequence of decisions, then the solution (what those decisions should be to achieve the extreme of some function) can often be found with *dynamic programming*. The development and proselytizing of this mathematical theory for solving multistage decision processes is most closely associated with the American mathematician Richard Bellman (1920–84). His 1957 book *Dynamic Programming* (Princeton) is recognized today as a classic and, although now almost a half-century old, it is still a source of fascinating problems.

Dynamic programming is much more of an "art form" than is linear programming, and in that sense resembles classical analysis much more than does linear programming. Indeed, the simplex algorithm is available in a number of different (huge) standardized computer programs into which one need only enter the objective function and constraint inequalities and out comes the answer. In dynamic programming, by contrast, we must develop a new analysis for each new problem, which generally takes the form of deriving a *functional equation* characteristic of the particular problem. Here's a simple but instructive example of that process for solving a problem, which will also illustrate the use of Bellman's famous *principle of optimality*:

If $P = x_1 x_2 \cdots x_n$, with $x_i \geq 0$ for $i = 1, 2, \cdots, n$, and if $\sum_{i=1}^{n} x_i = a$, a given constant, then what values for the x_i maximize P?

It should be obvious that, for $n \geq 2$, none of the x_i can be either 0 or a, as then $P = 0$, which is clearly not the largest P possible.

Whatever the answer is, it is certainly the case that it could only be a function (at most) of n—the *dimension* of the problem—and of a, as those are the only parameters in the problem. So, let's write the maximum value of P as $f_n(a)$. Now, for the trivial case of $n = 1$, i.e., for

$$P = x_1, x_1 \geq 0, \qquad \sum_{i=1}^{1} x_i = a = x_1,$$

we obviously have $f_1(a) = a$. For the somewhat more interesting case of $n = 2$, i.e., for

$$P = x_1 x_2, \qquad x_1 \geq 0, \qquad x_2 \geq 0, \qquad \sum_{i=1}^{2} x_i = a = x_1 + x_2,$$

we have $x_2 = a - x_1$ and so $P = x_1(a - x_1)$, where $0 \leq x_1 \leq a$. It is easy to see that to maximize P we should pick $x_1 = \frac{1}{2}a$ (and so $x_2 = \frac{1}{2}a$, as well). This gives $f_2(a) = \frac{1}{4}a^2$. How to proceed for the cases of $n \geq 3$, to find $f_3(a)$, $f_4(a)$, etc., is perhaps not so immediately clear, however. Consider the following approach.

Imagine that we have somehow found the value for x_1. We are left, then, with the problem of maximizing the product $x_2 x_3 \cdots x_n$ subject to the constraints

$$x_i \geq 0, \qquad i = 2, 3, \cdots, n$$

$$\sum_{i=2}^{n} x_i = a - x_1.$$

But that is exactly our *original problem*, except that we have reduced the problem dimension by 1 (from $n x_i$ to $n-1 x_i$), and that the $n-1 x_i$ sum to $a - x_1$ (not to a). Thus, by the very definition of f_n we can write the maximum value of the product $x_2 x_3 \cdots x_n$ as $f_{n-1}(a - x_1)$. So, to find the maximum value of P for the original problem, we pick x_1 to maximize $x_1 f_{n-1}(a - x_1)$, i.e., we have the multiplicative recurrence

$$f_n(a) = \max_{0 \leq x_1 \leq a} \{x_1 f_{n-1}(a - x_1)\},$$

which is the characteristic functional equation for this problem.

Observe, *carefully*, how we got this functional equation. We argued that the proper choices for the n-dimensional problem (the values for $x_1, x_2, \cdots, x_n$), what Bellman called the *optimal policy*, are such that the values of $x_2, x_3, \cdots, x_n$ form the optimal policy for the $(n-1)$-dimensional problem. This is Bellman's principle of optimality for a multistage decision process, which he stated as follows in his *Dynamic Programming*: "An optimal policy has the property that whatever the initial state and initial decision are [in our case, here, that is the value of x_1], the remaining decisions [that is, the values of $x_2, x_3, \cdots, x_n$] must constitute an optimal policy with regard to the state resulting from the first decision."

Once we have the functional equation for a problem, the second phase of a dynamic programming analysis is to *solve* (either analytically or, more usually, numerically) the functional equation. For our problem at hand, here's how to do that analytically. (I'll show you a numerical solution in my next example.) Let's return to the case of $n = 3$, the first case for which we found the problem getting less easy to handle. We have, from the functional equation, that

$$f_3(a) = \max_{0 \le x_1 \le a} \{x_1 f_2(a - x_1)\}.$$

But since $f_2(a) = \frac{1}{4}a^2$, then $f_2(a - x_1) = \frac{1}{4}(a - x_1)^2$, and so

$$f_3(a) = \max_{0 \le x_1 \le a} \left\{ x_1 \frac{1}{4}(a - x_1)^2 \right\}.$$

We can easily find the appropriate value of x_1 by simply setting the derivative of $\frac{1}{4}x_1(a - x_1)^2$ equal to zero. This gives $x_1 = \frac{1}{3}a$, and so

$$f_3(a) = \frac{a^3}{27}.$$

If we summarize our results so far, we have

$$f_1(a) = a = \left(\frac{a}{1}\right)^1,$$

$$f_2(a) = \frac{1}{4}a^2 = \left(\frac{a}{2}\right)^2,$$

$$f_3(a) = \frac{1}{27}a^3 = \left(\frac{a}{3}\right)^3.$$

It would seem to be obvious that a good guess to the general answer is

$$f_k(a) = \left(\frac{a}{k}\right)^k,$$

and this is, indeed, easy to confirm by induction. We certainly know the conjecture is true for $k = 1, 2$, and 3. So, let's assume the conjecture is true for $k = n - 1$ and see if that implies it is true for $k = n$. Thus,

$$f_n(a) = \max_{0 \le x_1 \le a} \{x_1 f_{n-1}(a - x_1)\} = \max_{0 \le x_1 \le a} \left\{ x_1 \left(\frac{a - x_1}{n - 1}\right)^{n-1} \right\}.$$

Setting the derivative of $x_1((a - x_1)/(n - 1))^{n-1}$ equal to zero gives $x_1 = a/n$, and so

$$f_n(a) = \frac{a}{n}\left(\frac{a - \dfrac{a}{n}}{n-1}\right)^{n-1} = \frac{a}{n}\left[\frac{na - a}{n(n-1)}\right]^{n-1} = \frac{a}{n}\left[\frac{a(n-1)}{n(n-1)}\right]^{n-1}$$

$$= \frac{a}{n}\left(\frac{a}{n}\right)^{n-1} = \left(\frac{a}{n}\right)^n,$$

which confirms the conjecture. Thus, to maximize P, set $x_1 = x_2 = \cdots = x_n = a/n$, which gives the maximum value of $P = x_1 x_2 \cdots x_n$ as $f_n(a) = (a/n)^n$.

In slightly less formal language, the principle of optimality is the mathematical version of what your parents always told you is the way to live an honorable life—"always do your best." That is, you'll make the most of what you started with if you always make the most of what you have left. Bellman was very much taken with the principle of optimality and, in his fascinating, eccentric autobiography *Eye of the Hurricane* (World Scientific), published the same year as his death (as befits a *true* autobiography), he wrote: "My first task in dynamic programming was to put it on a rigorous basis. I found that I was using the same technique over and over to derive a functional equation. I decided to call this technique, 'The principle of optimality.'"

When a friend objected, saying, "The principle is not rigorous," Bellman wrote that he replied " 'Of course not. It's not even precise.' A good principle should guide the intuition." As you might gather from this, Bellman was a character! He was a brilliant (even though his greatest admirers also thought him supremely arrogant) if somewhat erratic genius, and his job title at the time of his death shows the broad range of his interests: he held a professorship at the University of Southern California with joint appointments in mathematics, electrical engineering, biomedical engineering, and medicine. His breadth of mathematical accomplishment is illustrated by the recognition he received in 1979 from the world's largest *engineering* professional society, the Institute of Electrical and

Electronics Engineers (IEEE). That year Bellman received the IEEE Medal of Honor (the most prestigious award of all in electrical engineering) for his work in dynamic programming.

We are now in a position to understand how to use the principle of optimality to solve the directed-graph minimum-total-cost path problem from section 7.2; i.e., look back at figure 7.11. Let's write $c(i, j)$ as the cost of traveling the arc that links node i to node j (where you'll recall that our convention in numbering the nodes of a directed graph is that $i < j$). Also, let's write $f(i)$ as the *minimum* total cost in traveling from node i to the terminal node (node 8 in figure 7.11). Obviously, then, $f(8) = 0$.

Now, suppose that we have somehow arrived at node 7. There is only one way to travel to node 8, at a cost of $c(7, 8) = 4$. So, next to node 7 let's write a 4 inside of a box, as shown in figure 7.15. Similarly, if we have somehow arrived at either node 5 or node 6, then, in each case, there is only one way to get to node 8 (at the costs of $c(5, 8) = 1$ and $c(6, 8) = 1$, respectively). So, next to nodes 5 and 6 we write a 1 inside of a box. The numbers in the boxes represent the total cost to travel from a box's node to node 8, i.e., $f(7) = 4$, $f(6) = 1$, $f(5) = 1$.

Continuing to work our way backwards toward node 1 (it is, after all, the value of $f(1)$ and the path that achieve that minimum total cost that we are after), suppose next that somehow we have arrived at node 4. From there we now have *more* than one way to proceed; we could go to either node 5 or to node 6. If we go to node 5, the total cost to travel to node 8 is $c(4, 5) + f(5) = 4 + 1 = 5$, and if we go to node 6, the total cost to travel to node 8 is $c(4, 6) + f(6) = 1 + 1 = 2$. Since $f(4)$ is the *minimum* total cost to travel from node 4 to node 8, we obviously have

$$f(4) = \min \left\{ \begin{array}{l} c(4, 5) + f(5) \\ c(4, 6) + f(6) \end{array} \right\} = \min \left\{ \begin{array}{l} 5 \\ 2 \end{array} \right\} = 2.$$

So, next to node 4 we write 2 in a box. What we've done so far is shown in figure 7.15, which also shows a slanted bar struck through each arc that is traveled in going from one node to the next (this is equivalent to dropping rice behind us, so we can find our way back).

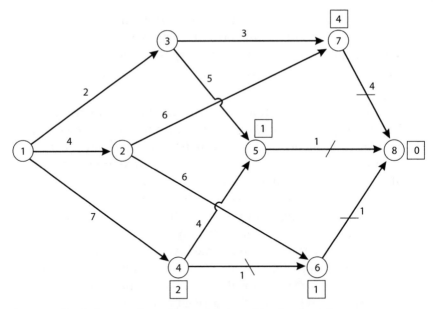

FIGURE 7.15. Solving a directed graph by working backwards.

The general process should now be clear. If we have somehow managed to arrive at node i, then the minimum total cost of traveling from that node to node 8 is given by

$$f(i) = \min_{j>i}\{c(i, j) + f(j)\},$$

and $f(i)$ is the number we write in a box next to node i. This additive recurrence is the dynamic programming functional equation for the least-cost path through a directed graph. It clearly incorporates the principle of optimality because, no matter how we may have gotten to a particular node, the path onward from that node to the terminal node is an optimal policy itself. If we continue to use the functional equation, we find that

$$f(3) = \min \left\{ \begin{matrix} c(3, 5) + f(5) \\ c(3, 7) + f(7) \end{matrix} \right\} = \min \left\{ \begin{matrix} 5 + 1 \\ 3 + 4 \end{matrix} \right\} = 6,$$

$$f(2) = \min \left\{ \begin{matrix} c(2, 7) + f(7) \\ c(2, 6) + f(6) \end{matrix} \right\} = \min \left\{ \begin{matrix} 6 + 4 \\ 6 + 1 \end{matrix} \right\} = 7,$$

$$f(1) = \min \left\{ \begin{array}{l} c(1,3) + f(3) \\ c(1,2) + f(2) \\ c(1,4) + f(4) \end{array} \right\} = \min \left\{ \begin{array}{l} 2+6 \\ 4+7 \\ 7+2 \end{array} \right\} = 8,$$

and the final result is shown in figure 7.16. The minimum cost path is $1 \to 3 \to 5 \to 8$, and the cost of that path is 8. It should be evident by now that the functional equation for this problem would be very easy to code for automatic execution on a computer and, given the connection topology of even a very large (say, $n = 1{,}000$ nodes) directed graph, the least-cost path could be found quickly.

There is one curious aspect to our solution—we found the optimal policy in the *reverse* order from which it would be actually implemented. That is, we worked backwards from node 8 to node 1 to find the least-cost path from node 1 to node 8. Since our convention for numbering the nodes means that we will encounter ever increasing node numbers as we move forward in time during our journey from node 1 to node 8, then *increasing* node numbers are a measure of *increasing* time. So, our numerical solution of the functional equation is a *backwards*-in-time process. This sort of thing is a

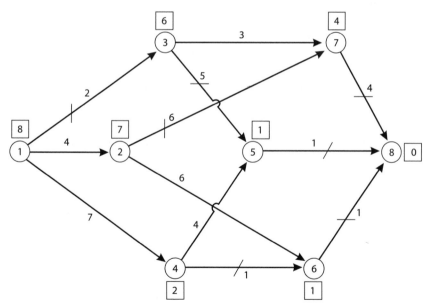

FIGURE 7.16. Final dynamic programming solution to the directed graph of figure 7.15.

general property of dynamic programming solutions, and it brings to mind an observation made more than 150 years ago by the Danish philosopher Søren Kierkegaard: "You can only understand life backwards, but we must live it forwards."

Our solution by dynamic programming of the directed graph problem is a pretty *mathematical* result, all by itself, but it has immediate practical applications, too. Consider, for example, the following problem faced by a manufacturing firm that sells a certain large, expensive earth-moving machine to construction engineers. Suppose we are given the following information about the firm's business practices and orders:

1. the construction of a machine takes one month;
2. if any machines *are* built during a month, then there is a fixed overhead production cost (independent of the number of machines constructed) of 2 units of money charged to the earth-moving budget (defining money this way keeps the numbers in the problem from becoming awkwardly large, e.g., define a *unit* to be $1,000);
3. the firm can construct any number of machines during any particular month (including none, with the production facility devoted that month to other manufacturing duties, and the overhead production cost charged against other budgets);
4. completed machines are shipped to customers only at the *end* of a month;
5. if a completed machine is not shipped at the end of a month, then it is stored on-site (as inventory) at a cost of 1 unit of money per machine per month (the month of construction does *not* incur a storage cost);
6. the production line for earth-moving machines is shut down during the winter months of December and January, and is available for production only during the other ten months of the year; the firm plans its operation in production cycles of 5-months duration, i.e., there are two production cycles per 12-month period.
7. at the start of each production cycle the inventory is zero;
8. the firm's marketing department has contracts for 16 machines, to be delivered according to the following schedule for the first production cycle:

At the end of	Number of machines to be shipped
February	2
March	4
April	2
May	5
June	3

The firm's production manager needs to determine the construction schedule (i.e., policy) that has the minimum total overhead/storage cost. That is, he needs to calculate how many machines should be constructed each month. One possible (extreme) policy, for example, would be to simply construct all 16 machines in February. The cost of doing that can be easily calculated as follows:

Machines Constructed during the Month of		Machines Stored during This Month	Machines Delivered at the End of This Month	Cost
February	16	0	2	2
March	0	14	4	14
April	0	10	2	10
May	0	8	5	8
June	0	3	3	3

The total overhead/storage cost for this particular policy is then 37 units of money. But is this the policy with the *minimum* total cost? The answer is no and, in fact, the optimal policy has a significantly lower cost.

To find the least-cost policy, we can represent the manager's problem as finding the least-cost path through a complete directed graph. To see how this is done, first observe that the following two conclusions can be made from the above list of the firm's business practices:

1. during any given month the firm should construct just enough machines to fill orders for an integer (including zero) number of months, because to do otherwise would result in excess machines that will incur storage costs while waiting for the next delivery date;

2. conclusion #1 implies the firm should construct new
   machines only when the inventory has shrunk to zero and,
   because of business practice #3, this causes no problems.

Now, let node i in a directed graph represent the firm at the *start* of
month i, where $i = 1 \Rightarrow$ February, $i = 2 \Rightarrow$ March, and so on, to
$i = 6 \Rightarrow$ July. (Notice that the *end* of June, when machines can last
be shipped, is the *start* of July.) This gives us the graph of figure 7.17,
where $c(i, j)$ is the cost of the arc joining node i to node j. That is,
$c(i, j)$ is the cost of, at the start of month i (with zero inventory),
constructing (and perhaps storing) enough machines to make all
deliveries to the start of month j. We have already, for example,
calculated that $c(1, 6) = 37$. Similar calculations result in the costs
shown in the figure (as shown in section 7.2, with $n = 6$ nodes there
are $\frac{1}{2}(6)(5) = 15$ arcs in this graph).

Applying the dynamic programming procedure for directed
graphs to the graph of figure 7.17 gives the result shown in figure
7.18, which shows that there are actually *two* production schedule

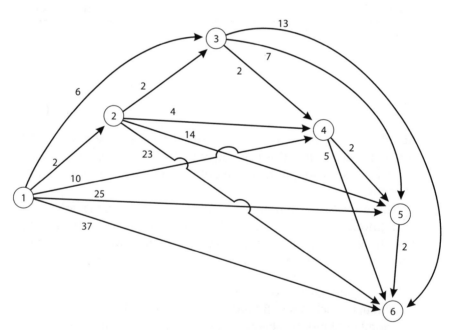

FIGURE 7.17. The directed graph of the machine construction problem.

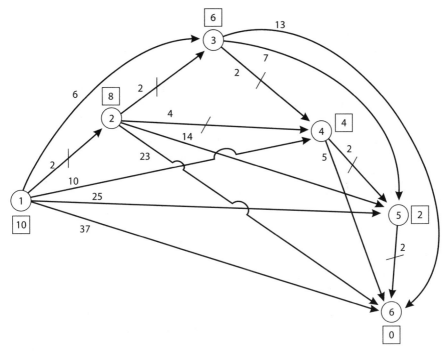

FIGURE 7.18. Dynamic programming solution of figure 7.17.

policies that result in the same least cost of 10 units of money: $1 \rightarrow 2 \rightarrow 3 \rightarrow 4 \rightarrow 5 \rightarrow 6$ and $1 \rightarrow 2 \rightarrow 4 \rightarrow 5 \rightarrow 6$. The optimal (least-cost) policy is not unique. The first policy says to construct 2 machines in February, 4 in March, 2 in April, 5 in May, and 3 in June, i.e., to construct the machines to be shipped at the end of each month during that month. The second policy says to construct 2 machines in February, 6 in March, none in April, 5 in May, and 3 in June. The choice between these two policies would have to be made on issues other than the cost, e.g., perhaps having the production facility available for another project during April would tip the decision toward the second policy.

A Question for You to Play With

Suppose the fixed overhead cost is changed to 4 units of money. The storage cost remains at 1 unit of money per machine per month. Now what is the optimal production policy? The answer is at the end of this section.

As the final example of dynamic programming, let's return to the 1,000-machine *integer* programming problem discussed in section 7.4. I left it unsolved there because the number of integer-valued variables (8) means the feasible solution set is a collection of *points* in hyperspace (which, as Dantzig observed, is a hard thing to "visualize"!) The dynamic programming formulation has no such complications. Recall that we are to determine how many of the still-operational machines to allocate, at the beginning of each week over a four-week period, to making part A (with the remaining operational machines assigned to making part B). As with the previous dynamic programming examples, we'll solve this problem "backwards in time."

To start, let's define $f_k(n)$ as the *maximum* profit that can be made during a period of k weeks that starts with n operational machines. So, if we assign x machines to make part A (and thus $n - x$ machines to make part B), then we can immediately write, for a one-week period $(k = 1)$,

$$f_1(n) = \max_{0 \leq x \leq n} \{400x + 600(n - x)\} = \max_{0 \leq x \leq n} (600n - 200x)$$

$$= 600n,$$

because we obviously achieve the maximum of $600n - 200x$ by setting $x = 0$. This result tells us that, at the *start* of week 4 (when we have just one week left), we should assign *all* of the then-operational machines to making to part B.

Let's now back up to the start of week 3, i.e., we are now concerned with $f_2(n)$, the maximum profit to be made from n operational machines with two weeks to go. Since $f_2(n)$ is the sum of the profits from week 3 and week 4, then if we start week 3 by assigning x machines to making part A (and so $n - x$ machines to making part B), we can write

$$f_2(n) = \max_{0 \leq x \leq n} \{[600n - 200x] + f_1[0.8x + 0.6(n - x)]\}.$$

Since $f_1(n) = 600n$, then

$$f_2(n) = \max_{0 \le x \le n} \{600n - 200x + 600[0.8x + 0.6(n - x)]\}$$

$$= \max_{0 \le x \le n} \{960n - 80x\} = 960n,$$

as we clearly maximize $960n - 80x$ by setting $x = 0$. So, as we start week 3, we should assign *all* of the operational machines to making part B.

Let's now back up to the start of week 2, i.e., we are now concerned with $f_3(n)$, the maximum profit to be made from n operational machines with three weeks to go. Since $f_3(n)$ is the sum of the profits from week 2 and the final two weeks, then if we start week 2 by assigning x machines to making part A (and so $n - x$ machines to making part B), we can write

$$f_3(n) = \max_{0 \le x \le n} \{[600n - 200x] + f_2[0.8x + 0.6(n - x)]\}.$$

Since $f_2(n) = 960n$, then

$$f_3(n) = \max_{0 \le x \le n} \{600n - 200x + 960[0.8x + 0.6(n - x)]\}$$

$$= \max_{0 \le x \le n} \{1{,}176n - 8x\} = 1{,}176n,$$

as we clearly maximize $1{,}176n - 8x$ by setting $x = 0$. So, as we start week 2, we should assign *all* of the operational machines to making part B.

Finally, let's back up to the start of week 1, i.e., we are now concerned with $f_4(n)$ (where of course now $n = 1{,}000$ for our particular problem). As usual, let's assign x machines to making part A and $n - x$ machines to making part B. Then, as before,

$$f_4(n) = \max_{0 \le x \le n} \{[600n - 200x] + f_3[0.8x + 0.6(n - x)]\},$$

or, as $f_3(n) = 1{,}176n$, we have

$$f_4(n) = \max_{0 \le x \le n} \{600n - 200x + 1{,}176[0.8x + 0.6(n - x)]\}$$

$$= \max_{0 \le x \le n} \{1{,}305.6n + 35.2x\} = 1340.8n,$$

as we clearly maximize $1{,}305.6n + 35.2x$ by setting $x = n$. So, as we start week 1, we should assign *all* of the operational machines to making part A.

Our optimal (maximum four-week profit) policy is, thus,

start of week 1: assign all 1,000 machines to making part A;
start of week 2: assign all remaining machines ($= 800$) to making part B;
start of week 3: assign all remaining machines ($= 480$) to making part B;
start of week 4: assign all remaining machines ($= 288$) to making part B.

This policy will generate a total profit of $1,340,800, and all other policies would generate *less* profit.

As a final comment on this problem, its solution by dynamic programming is in one important sense more general than is a solution by linear or integer programming; dynamic programming will still work even if the objective function is *non*linear. For example, suppose we keep all as before except for the profit that is generated from making each part. Suppose now that, instead of varying linearly with the amount of each part made (i.e., linearly with the number of machines assigned to make each part), the profit from each part varies *quadratically* with the number of machines assigned. This might occur, for example, from the reduced cost per part experienced by purchasing in large quantity the raw material required to make the parts. The problem of determining the optimal (maximum profit) policy is now one of *nonlinear programming* and the linear simplex algorithm will not work. Dynamic programming, however, doesn't miss a beat.

To see this, suppose that if n machines are assigned to make part A for a week, then the profit is $2n^2$, while if those n machines are assigned to make part B for a week, then the profit is $3n^2$. Defining $f_k(n)$ as before, if we have n operational machines at the start of week 4 and we assign x of them to making part A (and $n - x$ to making part B), then we can write

$$f_1(n) = \max_{0 \le x \le n} \left\{ 2x^2 + 3(n - x)^2 \right\} = \max_{0 \le x \le n} \left(5x^2 - 6nx + 3n^2 \right)$$

$$= 3n^2,$$

because $5x^2 - 6nx + 3n^2$ achieves its maximum value in the interval $0 \leq x \leq n$ at $x = 0$. This is because the quadratic expression has a *minimum* at an x *within* the interval (at $x = \frac{3}{5}n$), and so the maximum must occur at one of the endpoint values for x; which one is easy to determine. For $x = n$ the quadratic is $2n^2$, and for $x = 0$ the quadratic is $3n^2$. So, the conclusion is that $x = 0$. This result says that at the start of week 4 we should assign *all* operational machines to making part B, just as we concluded in the linear profit case.

Let's now back up to the start of week 3. Then, if we assign x machines out of n to making part A (and $n - x$ to making part B), we have

$$f_2(n) = \max_{0 \leq x \leq n} \left\{ 5x^2 - 6nx + 3n^2 + f_1[0.8x + 0.6(n - x)] \right\}$$

$$= \max_{0 \leq x \leq n} \left\{ 5x^2 - 6nx + 3n^2 + f_1(0.6n + 0.2x) \right\}$$

$$= \max_{0 \leq x \leq n} \left\{ 5x^2 - 6nx + 3n^2 + 3(0.6n + 0.2x)^2 \right\}$$

as $f_1(n) = 3n^2$. So,

$$f_2(n) = \max_{0 \leq x \leq n} \left\{ 5x^2 - 6nx + 3n^2 + 1.08n^2 + 0.72nx + 0.12x^2 \right\}$$

$$= \max_{0 \leq x \leq n} \left\{ 5.12x^2 - 5.28nx + 4.08n^2 \right\} = 4.08n^2$$

when $x = 0$. Thus, as we start week 3, we should assign all operational machines to making part B, just as we concluded in the linear profit case.

Let's now back up to the start of week 2. Then, as before, if we assign x out of n machines to making part A and $n - x$ to making part B, we have

$$f_3(n) = \max_{0 \leq x \leq n} \left[5x^2 - 6nx + 3n^2 + f_2[0.6n + 0.2x] \right\}$$

$$= \max_{0 \leq x \leq n} \left\{ 5x^2 - 6nx + 3n^2 + 4.08(0.6n + 0.2x)^2 \right\}$$

$$= \max_{0 \leq x \leq n} \left\{ 5.1632x^2 - 5.0208nx + 4.4688n^2 \right\}$$

$$= 4.6112n^2$$

when $x = n$. So, as we start week 2, we should assign all operational machines to making part A, which is *not* what the optimal policy says to do in the linear profit case.

And finally, let's back up to the start of week 1. Then,

$$f_4(n) = \max_{0 \leq x \leq n} \left\{ 5x^2 - 6nx + 3n^2 + f_3[0.6n + 0.2x] \right\}$$

$$= \max_{0 \leq x \leq n} \left\{ 5x^2 - 6nx + 3n^2 + 4.6112(0.6n + 0.2x)^2 \right\}$$

$$= \max_{0 \leq x \leq n} \left\{ 5.184448x^2 - 4.893312nx + 4.660032n^2 \right\}$$

$$= 4.951168n^2$$

when $x = n$. So, as we start week 1, we should assign all operational machines to making part A, just as we concluded in the linear profit case. The quadratic profit function has caused the decision at the start of week 2 in the optimal policy to switch from what it is in the linear profit case.

The next complication we might introduce in these calculations is to recognize that some of the given conditions of the problem are unrealistic. For example, why would *exactly* 20% (40%) of the machines making part A (part B) during each week break each and every week? On *average* this could perhaps make sense, but from week to week to week a better model would specify probability density functions for the breakdown percentages. But then the optimal policy solution would also have a probability density function and what could *that* mean? One possible answer is to say that the optimal policy is optimal *on the average*, i.e., if the multistage decision process is one that is repeated over and over many times, then the *average* cost is minimized by that policy. For a process that is to be carried out only once, however, this definition of optimality has no meaning. What do we do then? There *is* an answer to that question, too, but you are not going to find it here.

In the spirit of this book's title, more is *not* always better. And so, at last, this is finally *the end*.

Solution to the Challenge Problem

The directed graph for the modified earth-moving machine problem (with the fixed overhead cost changed to 4 units of

(continued)
money and the storage cost remaining at 1 unit of money per machine per month) is shown in figure 7.19, along with the dynamic programming results. The optimal policy is $1 \to 2 \to 4 \to 6$, with a cost of 17, i.e., *the* optimal production policy (now unique) is to construct 2 machines in February, 6 in March, none in April, 8 in May, and none in June.

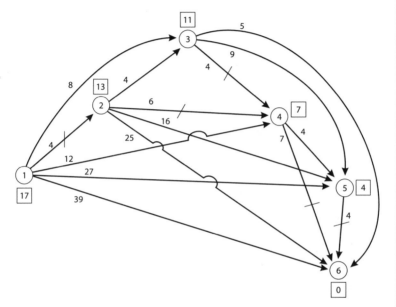

FIGURE 7.19. Dynamic programming solution to the challenge problem.

A final note: I opened this book with a number of quotes to indicate the importance of studying extrema. Let me close with one last quote, an (unintentionally hilarious) illustration of the elusive nature of extrema among even highly educated professionals. In a recent (2002) report on the need to sensitize doctors to the benefits of eliminating avoidable pain during medical procedures, the authors concluded that "Optimal pain control should be the minimum acceptable standard." No one would disagree in spirit with the noble nature of this goal, but I'm afraid it is a priori an impossible goal. After reading this book, you should know why.

Appendix A.

The AM-GM Inequality

If $x_1, x_2, \cdots, x_n$ are *any* n nonnegative numbers, $n \geq 1$, and if $A = (1/n)(x_1 + x_2 + \cdots + x_n)$ is the arithmetic mean of the x's, and if $G = (x_1 x_2 \cdots x_n)^{1/n}$ is the geometric mean of the x's, then $A \geq G$ with equality iff $x_1 = x_2 = \cdots = x_n$.

This was known to Euclid for the $n = 2$ case. The first proof, for arbitrary n, is due to the Scottish mathematician Colin Maclaurin (1698–1746), who published it in 1729.

PROOF. Suppose we have any n positive numbers (if one or more of the x_i are zero, then the inequality is trivially obvious, and so we'll suppose all the $x_i > 0$) whose product is 1, i.e.,

$$x_1 x_2 \cdots x_n = 1.$$

This may seem at odds with the above statement that the x's can be *any* n positive numbers, because the product may then *not* be 1. Suppose, in fact, the product is P. In that case we divide through both sides by P and replace each x_i with

$$y_i = \frac{x_i}{P^{1/n}}.$$

Then we *do* have

$$y_1 y_2 \cdots y_n = 1.$$

I'll continue on at this point with the y's and, at the end, simply replace each y_i with $x_i / P^{1/n}$. You'll see soon how nicely things then turn out.

Now, it is clear that either all of the y_i are equal (and so all are equal to 1) or that they are not all equal. If they are all equal (to 1), it is equally clear that then

$$y_1 + y_2 + \cdots + y_n = n.$$

If they are not all equal, then the claim is that their sum is *at least* equal to n, i.e.,

$$y_1 + y_2 + \cdots + y_n \geq n$$

with equality iff $y_1 = y_2 = \cdots = y_n$. Thus, our claim is that this last inequality is true whether all the y_i are equal or not (as obviously $n \geq n$). I'll establish this result in the next paragraph but, assuming for now this is so, you can see how the AM-GM inequality immediately follows. That's because if we divide through by n, we have

$$\frac{y_1 + y_2 + \cdots + y_n}{n} \geq 1 = y_1 y_2 \cdots y_n = 1^{1/n} = (y_1 y_2 \cdots y_n)^{1/n}.$$

Then, replacing each y_i with $x_i/P^{1/n}$ as explained above,

$$\frac{x_1 + x_2 + \cdots + x_n}{n P^{1/n}} \geq \left(\frac{x_1 x_2 \cdots x_n}{P}\right)^{1/n} = \frac{(x_1 x_2 \cdots x_n)^{1/n}}{P^{1/n}},$$

or, as P now conveniently vanishes from the inequality, we have

$$\frac{x_1 + x_2 + \cdots + x_n}{n} \geq (x_1 x_2 \cdots x_n)^{1/n},$$

with equality iff $x_1 = x_2 = \cdots = x_n$. So, all we have to do is show the truth of our assumption, i.e., that, indeed, $y_1 + y_2 + \cdots + y_n \geq n$.

We've already seen that the case of all the y_i equal (to 1) is trivial, so now we'll treat the case where all the y_i are not equal. Can all the y_i be greater than 1? No, as then their product would be greater than 1. Can all of the y_i be less than 1? Again, no, as then their product would be less than 1. So, at least one y_i must be greater than 1 (label it y_1) and at least one y_i must be less than 1 (label it y_2). Thus, $1 - y_1 < 0$ and $1 - y_2 > 0$, and so

$$(1 - y_1)(1 - y_2) < 0,$$

or

$$1 - y_1 - y_2 + y_1 y_2 < 0,$$

or

$$1 + y_1 y_2 < y_1 + y_2.$$

What are we going to do with this? We'll use it to complete an induction proof, i.e., we'll assume that our claim is true for $n = k$ (that $y_1 y_2 \cdots y_k = 1$ means $y_1 + y_2 + \cdots + y_k \geq k$) and then show that the claim must be true for $n = k + 1$. Since the claim is obviously true for $n = 1$, then the claim would be true for $n = 2$, and so on; it would be true for all $n \geq 1$.

So, by assumption we have $y_1 y_2 \cdots y_k = 1$ and $y_1 + y_2 + \cdots + y_k \geq k$, i.e., if any k positive numbers have a product of one then their sum is at least k. Our concern now is with $k + 1$ positive numbers whose product is 1—what can we say about their sum? I'll write these $k + 1$ numbers as $\hat{y}_i$, $1 \leq i \leq k+1$, to indicate that they are not necessarily the k y_i numbers. The only important consideration is that now we have $k + 1$ numbers. So, we have

$$\hat{y}_1 \hat{y}_2 \cdots \hat{y}_k \hat{y}_{k+1} = 1,$$

and we consider the sum $\hat{y}_1 + \hat{y}_2 + \cdots + \hat{y}_k + \hat{y}_{k+1}$. By the same argument above it must be true that

$$1 + \hat{y}_1 \hat{y}_2 < \hat{y}_1 + \hat{y}_2,$$

and so

$$\hat{y}_1 + \hat{y}_2 + \cdots + \hat{y}_k + \hat{y}_{k+1} \geq 1 + \hat{y}_1 \hat{y}_2 + \hat{y}_3 + \cdots + \hat{y}_{k+1}.$$

But, $\hat{y}_1 \hat{y}_2 + \hat{y}_3 + \cdots + \hat{y}_{k+1}$ is the sum of k (*not* $k+1$) positive numbers whose product is $\hat{y}_1 \hat{y}_2 \hat{y}_3 \cdots \hat{y}_{k+1} = 1$ and we already know that sum is at least k. So,

$$\hat{y}_1 + \hat{y}_2 + \cdots + \hat{y}_k + \hat{y}_{k+1} \geq 1 + k$$

and we are done.

Appendix B.
The AM-QM Inequality, and Jensen's Inequality

If $x_1, x_2, \cdots, x_n$ are any n real numbers, then

$$\frac{x_1 + x_2 + \cdots + x_n}{n} \leq \sqrt{\frac{x_1^2 + x_2^2 + \cdots + x_n^2}{n}}$$

with equality iff $x_1 = x_2 = \cdots = x_n$.

PROOF. Squaring the arithmetic mean, we have

$$\left(\frac{x_1 + x_2 + \cdots + x_n}{n}\right)^2$$

$$= \frac{x_1^2 + x_2^2 + \cdots + x_n^2 + \text{all possible } x_i x_j \text{ cross-products with } i \neq j}{n^2}.$$

We can write this more compactly (and more clearly, as well, I think) as

$$\left(\frac{x_1 + x_2 + \cdots + x_n}{n}\right)^2 = \left(\frac{\sum\limits_{i=1}^{n} x_i}{n}\right)^2 = \frac{\sum\limits_{i=1}^{n} x_i^2 + \sum\limits_{\substack{i=1 \\ i \neq j}}^{n} \sum\limits_{j=1}^{n} x_i x_j}{n^2}.$$

Now, since the square of a real number is never negative, we have

$$0 \leq \left(x_i - x_j\right)^2 = x_i^2 - 2x_i x_j + x_j^2,$$

and so $2x_i x_j \leq x_i^2 + x_j^2$ with equality iff $x_i = x_j$. Thus,

$$\left(\frac{\sum_{i=1}^{n} x_i}{n}\right)^2 \leq \frac{\sum_{i=1}^{n} x_i^2 + \frac{1}{2} \sum_{\substack{i=1 \\ i \neq j}}^{n} \sum_{j=1}^{n} \left(x_i^2 + x_j^2\right)}{n^2}$$

$$= \frac{\frac{1}{2} \sum_{i=1}^{n} \left(x_i^2 + x_i^2\right) + \frac{1}{2} \sum_{\substack{i=1 \\ i \neq j}}^{n} \sum_{j=1}^{n} \left(x_i^2 + x_j^2\right)}{n^2}.$$

This may look somewhat cryptic, but the right-hand side becomes easy to visualize if you sketch the $n \times n$ array (or *square matrix*) of values as shown in the table below, where the value of the element at (i, j) is $x_i^2 + x_j^2$. The first summation, $\sum_{i=1}^{n} (x_i^2 + x_i^2)$, is the sum of the entries along the main diagonal, while the second summation $\sum_{\substack{i=1 \\ i \neq j}}^{n} \sum_{j=1}^{n} (x_i^2 + x_j^2)$, is the sum of all the off-diagonal entries.

$$\underline{\quad\quad} j \longrightarrow$$

		1	2	3	4	·
	1	$x_1^2 + x_1^2$	$x_1^2 + x_2^2$	$x_1^2 + x_3^2$	$x_1^2 + x_4^2$	etc.
i	2	$x_2^2 + x_1^2$	$x_2^2 + x_2^2$	$x_2^2 + x_3^2$	$x_2^2 + x_4^2$	etc.
	3	$x_3^2 + x_1^2$	$x_3^2 + x_2^2$	$x_3^2 + x_3^2$	$x_3^2 + x_4^2$	etc.
	4	$x_4^2 + x_1^2$	$x_4^2 + x_2^2$	$x_4^2 + x_3^2$	$x_4^2 + x_4^2$	etc.
	·	etc.	etc.	etc.	etc.	etc.
	·	etc.				
	·					

Thus, the sum of the two summations is simply the sum of *all* of the terms in the array, i.e., our inequality becomes

$$\left(\frac{\sum_{i=1}^{n} x_i}{n}\right)^2 \leq \frac{\frac{1}{2} \sum_{i=1}^{n} \sum_{j=1}^{n} \left(x_i^2 + x_j^2\right)}{n^2}.$$

The table clearly shows that each x^2 term, for a given subscript, appears a total of $2n$ times in the array, and so

$$\frac{\left(\sum\limits_{i=1}^{n} x_i\right)^2}{n} \leq \frac{\frac{1}{2} 2n \sum\limits_{i=1}^{n} x_i^2}{n^2} = \frac{1}{n} \sum\limits_{i=1}^{n} x_i^2,$$

with equality iff $x_1 = x_2 = \cdots = x_n$. The right-hand side is the *quadratic mean* (QM). Taking the square root, we have

$$\frac{1}{n} \sum\limits_{i=1}^{n} x_i \leq \sqrt{\frac{1}{n} \sum\limits_{i=1}^{n} x_i^2}$$

with equality iff $x_1 = x_2 = \cdots = x_n$, and we are done. The right-hand side of this inequality is often called the rms value of the x's (rms is the abbreviation for "root-mean-square") as it is the square *root* of the *mean* of the *squares*.

The AM-QM inequality is actually just a special case of a far more general result called *Jensen's inequality* (see section 2.2 for who Jensen was). To understand Jensen's result, let's start by considering the graph of the function $f(x) = x^2$ in figure B1, which is of course an upward-opening parabola. If we take *any* two values of x, say x_1 and x_2, then it is geometrically clear that the chord joining the two points on the parabola (x_1, x_1^2) and (x_2, x_2^2) lies *above* the section of the parabola cut off by the chord. This property identifies $f(x) = x^2$ as what mathematicians call a *strictly convex* function. If, on the other hand, $f(x)$ is a function whose graph opens *downward* (e.g., $f(x) = \sin(x)$ for $0 \leq x \leq \pi$), then the chord joining any two points on $f(x)$ lies *below* the curve and $f(x) = \sin(x)$ is said to be strictly *concave* over the interval $0 \leq x \leq \pi$.

For such functions, Jensen's inequality says:

if $f(x)$ is strictly convex (strictly concave) on some interval, and if $x_1, x_2, \cdots, x_n$ are n values of x from that interval, and if $c_1, c_2, \cdots, c_n$ are n positive constants such that $c_1 + c_2 + \cdots + c_n = 1$, then

$$f\left(\sum\limits_{i=1}^{n} c_i x_i\right) \underset{\leq}{\overset{\geq}{=}} \sum\limits_{i=1}^{n} c_i f(x_i)$$

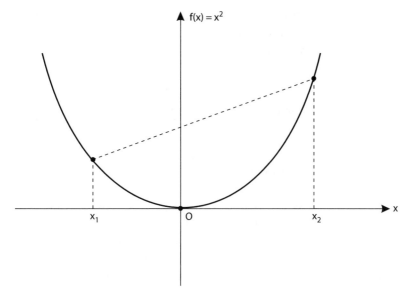

FIGURE B1. A parabola is a strictly convex function.

with equality iff $x_1 = x_2 = \cdots = x_n$, where we use $\leq$ if $f(x)$ is convex, and we use $\geq$ if $f(x)$ is concave.

For example, it is geometrically clear that $f(x) = x^2$ is strictly convex on the entire real line and so, if we pick $c_1 = c_2 = \cdots = c_n = 1/n$, then Jensen's inequality becomes for *any* set of n numbers $x_1, x_2, \cdots, x_n$,

$$f\left(\frac{1}{n}\sum_{i=1}^{n} x_i\right) \leq \frac{1}{n}\sum_{i=1}^{n} f(x_i),$$

or

$$\left(\frac{x_1 + x_2 + \cdots + x_n}{n}\right)^2 \leq \frac{x_1^2 + x_2^2 + \cdots + x_n^2}{n},$$

or

$$\frac{x_1 + x_2 + \cdots + x_n}{n} \leq \sqrt{\frac{x_1^2 + x_2^2 + \cdots + x_n^2}{n}}$$

with equality iff $x_1 = x_2 = \cdots = x_n$, which is the AM-QM inequality.

As another example, if $f(\theta) = \sin(\theta)$, which is *strictly concave* in the interval $0 \leq \theta \leq \pi$, and if we pick $c_1 = c_2 = \cdots = c_n = 1/n$, then Jensen's inequality becomes for any set of n numbers in the interval 0 to π,

$$\sin\left(\frac{1}{n}\sum_{i=1}^{n}\theta_i\right) \geq \frac{1}{n}\sum_{i=1}^{n}\sin(\theta_i),$$

with equality iff $\theta_1 = \theta_2 = \cdots = \theta_n$, a result used in chapter 2 to show that the maximum area N-gon inscribed in a given circle is a *regular N-gon*.

And finally, as you probably suspect by now, we can derive the AM-GM inequality (which started all of our discussion of inequalities) as also simply a special case of Jensen's inequality. If we pick $f(x) = -\ln(x)$, a function that is strictly convex over the *nonnegative* real axis ($f(x)$ is complex for $x < 0$), then Jensen's inequality becomes

$$-\ln\left(\sum_{i=1}^{n}c_i x_i\right) \leq -\sum_{i=1}^{n}c_i \ln(x_i) \qquad \text{for } x_i \geq 0,$$

with equality iff $x_1 = x_2 = \cdots = x_n$. That is,

$$\ln\left(c_1 x_1 + c_2 x_2 + \cdots + c_n x_n\right) \geq \ln\left(x_1^{c_1}\right) + \ln\left(x_2^{c_2}\right) + \cdots + \ln\left(x_n^{c_n}\right)$$

$$= \ln\left(x_1^{c_1} x_2^{c_2} \cdots x_n^{c_n}\right).$$

Thus,

$$c_1 x_1 + c_2 x_2 + \cdots + c_n x_n \geq x_1^{c_1} x_2^{c_2} \cdots x_n^{c_n},$$

and so, if we pick $c_1 = c_2 = \cdots = c_n = 1/n$, then, with all the $x_i > 0$, we have

$$\frac{x_1 + x_2 + \cdots + x_n}{n} \geq (x_1 x_2 \cdots x_n)^{1/n},$$

with equality iff $x_1 = x_2 = \cdots = x_n$, which is the AM-GM inequality.

Clearly, Jensen's inequality is a very powerful result and deserves to be better known than it is, at least among engineers and scientists, especially considering that Jensen himself was an engineer! It is also not difficult to prove; I'll do it here for the case of $f(x)$ strictly convex, and the strictly concave version will then follow immediately. That is so because if $f(x)$ is strictly concave, then $-f(x)$ is strictly convex and, after multiplying through the inequality for $-f(x)$ by minus one, all that happens is that the sense of the inequality is reversed.

Proof (by induction) of Jensen's Inequality.

Step 1. The inequality is true for the case of $n = 2$ and $f(x)$ convex because it says

$$f(c_1 x_1 + c_2 x_2) \le c_1 f(x_1) + c_2 f(x_2), \qquad c_1 + c_2 = 1,$$

with equality iff $x_1 = x_2$. That is, if we drop the subscripts and simply write c for c_1 and $1 - c$ for c_2, the inequality says (for $n = 2$)

$$f[c x_1 + (1 - c)x_2] \le c f(x_1) + (1 - c)f(x_2), \qquad 0 < c < 1.$$

But this is just what is meant *geometrically* by saying $f(x)$ is strictly convex; the left-hand side is the height of the plot of f at an arbitrary value of x between $x = x_1$ and $x = x_2$ (x varies from x_1 to x_2 as c varies from 1 to 0), while the right-hand side is the height of the *straight-line* chord joining the two points $(x_1, f(x_1))$ and $(x_2, f(x_2))$, which varies *linearly* between $f(x_1)$ and $f(x_2)$ as c varies between 1 and 0. That is, for $n = 2$, the inequality simply says that the chord lies *above* the graph of the function for any x such that $x_1 < x < x_2$.

To show the iff condition, first suppose $x_1 = x_2$. Then, the inequality says

$$f[c x_1 + (1 - c)x_1] \le c f(x_1) + (1 - c)f(x_1).$$

Thus, as *both* sides reduce to $f(x_1)$, we have *equality*. To go in the opposite direction, now suppose that

$$f[c x_1 + (1 - c)x_2] = c f(x_1) + (1 - c)f(x_2), \qquad 0 < c < 1.$$

But this says the *function* and the *chord* are equal at all points between x_1 and x_2 which is clearly impossible for a strictly convex f *unless* $x_1 = x_2$. This shows the iff condition for the $n = 2$ case.

Step 2. We now assume that the inequality is true for $n = k$ and show that this implies it is true for $n = k + 1$. That is, we take as true

$$f\left(\sum_{i=1}^{k} c_i x_i\right) \leq \sum_{i=1}^{k} c_i f(x_1) \qquad \text{with } \sum_{i=1}^{k} c_i = 1, \qquad c_i > 0$$

and ask what we can say about

$$f\left(\sum_{i=1}^{k+1} \hat{c}_i x_i\right) \qquad \text{with } \sum_{i=1}^{k+1} \hat{c}_i = 1, \qquad \hat{c}_i > 0?$$

Notice that I am *not* assuming the first k values of the $k+1\hat{c}_i$ are the kc_i; only that whatever the $\hat{c}_i$ are they satisfy the conditions of all being positive, and summing to 1.

Now, we can write

$$f\left(\sum_{i=1}^{k+1} \hat{c}_i x_i\right) = f\left[\left\{(1 - \hat{c}_{k+1})\sum_{i=1}^{k} \frac{\hat{c}_i}{1 - \hat{c}_{k+1}} x_i\right\} + \hat{c}_{k+1} x_{k+1}\right].$$

From Step 1, we have, by the *definition* of strict convexity of f, that

$$f[(1 - c)u + cv] \leq cf(v) + (1 - c)f(u),$$

and so

$$f\left(\sum_{i=1}^{k+1} \hat{c}_i x_i\right) \leq \hat{c}_{k+1} f(x_{k+1}) + (1 - \hat{c}_{k+1}) f\left(\sum_{i=1}^{k} \frac{\hat{c}_i}{1 - \hat{c}_{k+1}} x_i\right).$$

Now, since $\hat{c}_i/(1 - \hat{c}_{k+1}) > 0$ for all i (because $\hat{c}_i > 0$ for all i, and $\hat{c}_{k+1} < 1$ because the $\hat{c}_i$ sum to 1), and since

$$\sum_{i=1}^{k} \frac{\hat{c}_i}{1 - \hat{c}_{k+1}} = \frac{\hat{c}_1}{1 - \hat{c}_{k+1}} + \frac{\hat{c}_2}{1 - \hat{c}_{k+1}} + \cdots + \frac{\hat{c}_k}{1 - \hat{c}_{k+1}}$$

$$= \frac{\hat{c}_1 + \hat{c}_2 + \cdots + \hat{c}_k}{1 - \hat{c}_{k+1}} = \frac{1 - \hat{c}_{k+1}}{1 - \hat{c}_{k+1}} = 1,$$

then from the assumed truth of the $n = k$ case, we have

$$f\left(\sum_{i=1}^{k} \frac{\hat{c}_i}{1 - \hat{c}_{k+1}} x_i\right) \leq \sum_{i=1}^{k} \frac{\hat{c}_i}{1 - \hat{c}_{k+1}} f(x_i).$$

But this says

$$f\left(\sum_{i=1}^{k+1} \hat{c}_i\, x_i\right) \le \hat{c}_{k+1} f(x_{k+1}) + (1 - \hat{c}_{k+1}) \sum_{i=1}^{k} \frac{\hat{c}_i}{1 - \hat{c}_{k+1}}\, f(x_i)$$

$$= \hat{c}_{k+1} f(x_{k+1}) + \sum_{i=1}^{k} \hat{c}_i\, f(x_i)$$

$$= \sum_{i=1}^{k+1} \hat{c}_i f(x_i).$$

That is, the truth of the inequality for the $n = k+1$ case follows from the assumption the inequality holds for the $n = k$ case. And, since the inequality *does* hold for the $n = 2$ case, it holds for *all $n > 2$* as well, and we are done. I'll let *you* fill in the remaining iff arguments.

As a final note, Jensen's inequality (1906) was actually derived earlier (1889) by the German mathematician Otto Hölder (1859–1937), but in a formal, nongeometric context, i.e., Hölder's initial assumption was simply that the second derivative of $f(x)$ exist and be nonnegative. The *geometric* interpretation of a convex function is Jensen's (as is the term *convex*). Another famous inequality *is* named after Hölder (1884), but it is not the one studied here.

Appendix C.
"The Sagacity of the Bees" (the preface to Book 5 of Pappus' *Mathematical Collection*)

Though God has given to men, most excellent Megethion, the best and most perfect understanding of wisdom and mathematics, He has allotted a partial share to some of the unreasoning creatures as well. To men, as being endowed with reason, He granted that they should do everything in the light of reason and demonstration, but to the other unreasoning creatures He gave only this gift, that each of them should, in accordance with a certain natural forethought, obtain so much as is needful for supporting life. This instinct may be observed to exist in many other species of creatures, but it is specially marked among bees. Their good order and their obedience to the queens who rule in their commonwealths are truly admirable, but much more admirable still is their emulation, their cleanliness in the gathering of honey, and the forethought and domestic care they give to its protection. Believing themselves, no doubt, to be entrusted with the task of bringing from the gods to the more cultured part of mankind a share of ambrosia in this form, they do not think it proper to pour it carelessly into earth or wood or any other unseemly and irregular material, but, collecting the fairest parts of the sweetest flowers growing on the earth, from them they prepare for the reception of the honey the vessels called honeycombs, [with cells] all equal, similar and adjacent, and hexagonal in form.

That they have contrived this in accordance with a certain geometrical forethought we may thus infer. They would necessarily
think that the figures must all be adjacent one to another and have
their sides common, in order that nothing else might fall into the
interstices and so defile their work. Now there are only three rec-
tilineal figures which would satisfy the condition, I mean regular
figures which are equilateral and equiangular, inasmuch as irregular figures would be displeasing to the bees. For equilateral triangles and squares and hexagons can lie adjacent to one another
and have their sides in common without irregular interstices. For
the space about the same point can be filled by six equilateral triangles and six angles, of which each is $\frac{2}{3}$ [of a] right angle, or by
four squares and four right angles, or by three hexagons and three
angles of a hexagon, of which each is $1\frac{1}{3}$ [of a] right angle. But
three pentagons would not suffice to fill the space about the same
point, and four would be more than sufficient; for three angles
of the pentagon are less than four right angles (inasmuch as each
angle is $1\frac{1}{5}$ [of a] right angle), and four angles are greater than four
right angles. Nor can three heptagons be placed about the same
point so as to have their sides adjacent to each other; for three
angles of a heptagon are greater than four right angles (inasmuch
as each is $1\frac{3}{7}$ [of a] right angle). And the same argument can be
applied even more to polygons with a greater number of angles.
There being, then, three figures capable by themselves of filling
up the space around the same point, the triangle, the square and
the hexagon, the bees in their wisdom chose for their work that
which has the most angles, perceiving that it would hold more
honey than either of the two others.

Bees, then, know just this fact which is useful to them, that the
hexagon is greater than the square and the triangle and will hold
more honey for the same expenditure of material in constructing
each. But we, claiming a greater share in wisdom than the bees,
will investigate a somewhat wider problem, namely that, *of all
equilateral and equiangular plane figures having an equal perimeter,
that which has the greater number of angles is always greater, and the
greatest of them all is the circle having its perimeter equal to them.*

Pappus' ancient words motivated mathematicians many centuries
later, by then in possession of the calculus, to analytically study the

"best" way to make a honeycomb. Two such mathematicians were the German Johann Samuel König (1712–57), who published his analysis (with some errors) in 1740, and then later (1755) Ruggero Boscovich (1711–87). You can find Boscovich's work described in the paper by R. M. Dimitrić: "Using Less Calculus in Teaching Calculus: An Historical Approach" (*Mathematics Magazine*, June 2001, pp. 201–11). For what a modern mathematician has to say on just how well the bees actually do, see L. Fejes Tóth, "What the Bees Know and What They Do Not Know" (*Bulletin of the American Mathematical Society*, 1964, pp. 468–81). Tóth concludes that they do pretty well! As he wrote, "We must admit that all this [i.e., Tóth's construction of a honeycomb structure *just slightly* more efficient than the structures real bees actually build] has no practical consequence. By building such cells [Tóth's cells] the bees would save per cell less than 0.35% of the area of an opening . . . under such conditions the above 'saving' is quite illusory. Besides, the building style of bees is definitely simpler . . . so we would fail in shaking someone's conviction that the bees have a deep geometrical intuition."

Appendix D.

Every Convex Figure Has
a Perimeter Bisector

Let φ denote a convex figure, and Q a point not inside or on the boundary edge of φ, as shown in figure D1. Let a line be drawn through Q, with α denoting the angle that line makes with the x-axis. In the figure I've assumed that φ is positioned so that it lies entirely in the first quadrant, above the positive x-axis and to the right of the positive y-axis. To make things really easy to visualize

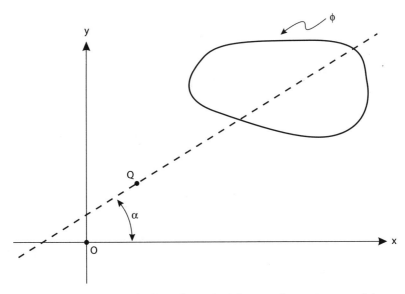

FIGURE D1. For some α, the line through Q bisects the perimeter of ϕ.

and explain, I have also assumed that φ is positioned so that the left- and bottommost points of φ are to the right of and above Q, respectively, as shown in figure D1. None of these assumptions limits the generality of our eventual result. Finally, let $P(\alpha)$ denote the fraction of φ's perimeter *below* the line we drew through Q. Thus $0 \leq P(\alpha) \leq 1$ with $P(0°) = 0$ with $P(90°) = 1$.

It is geometrically clear that $P(\alpha)$ is a *smoothly* increasing function of increasing α, i.e., $P(\alpha)$ is what mathematicians call a *continuous* function (there are no sudden, discontinuous jumps in the value of $P(\alpha)$ as α increases from $0°$ to $90°$). So, in particular, there must be some $\alpha = \hat{\alpha}$ where $P(\hat{\alpha}) = \frac{1}{2}$, i.e., at angle $\alpha = \hat{\alpha}$, the line through Q bisects the perimeter of φ. Notice that this is an existence proof; it tells us only, for a given φ and Q, that there *is* a line at some angle $\hat{\alpha}$ that bisects the perimeter of φ, but it does not tell us what $\hat{\alpha}$ is. It should also now be clear, since we could locate Q in infinitely many places, that there is not just *one* perimeter bisector for φ but, in fact, there is an infinity of perimeter bisectors.

Appendix E.
The Gravitational Free-Fall Descent Time along a Circle

The exact analysis of Galileo's problem, that of determining the descent time of a bead, due to gravity, constrained to a vertical circular path of radius L, is a classic in the marriage of physics and calculus. In figure E1, a bead of mass m is constrained to move along a vertical circular wire arc that threads through a hole in the bead. It is assumed that friction can be ignored. The initial angle the radius to the bead makes with the vertical radius is α, and the instantaneous angle, as the bead slides along the wire is θ (also measured with respect to the vertical radius). That is, $\theta(t = 0) = \alpha$. If the bead arrives at the bottom of the wire at time $t = T$, then of course $\theta(t = T) = 0$. Our problem here is to calculate T.

At time t let the distance *along the circular path* from the bead to the bottom of the wire be s, and let $v(t)$ be the speed of the bead. Then, since we are ignoring friction, we can set the bead's change in kinetic energy of motion equal to the change in its potential energy of position. We assume that the bead starts its fall from rest, i.e., that $v(0) = 0$. Then, at time t, and writing g for the acceleration of gravity, we have

$$\frac{1}{2}mv^2 = mg\{L\cos(\theta) - L\cos(\alpha)\}.$$

Also, since $s = L\theta$ and as $v = ds/dt$, then

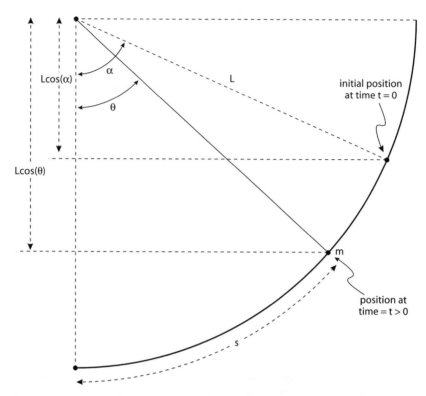

FIGURE E1. A bead sliding under gravity along a vertical, circular wire.

$$v = L \frac{d\theta}{dt}$$

and, thus,

$$\frac{1}{2} L^2 \left(\frac{d\theta}{dt} \right)^2 = gL \left\{ \cos(\theta) - \cos(\alpha) \right\}.$$

We can now take great advantage of Leibniz's differential notation, treat the differential dt just as an algebraic quantity, and solve
for it. To start, write

$$\frac{d\theta}{dt} = \pm \sqrt{\frac{2g}{L} \left\{ \cos(\theta) - \cos(\alpha) \right\}}.$$

Since θ *decreases* as t (time) increases (because the bead is sliding downward) we know the change in θ has algebraic sign opposite to that of the change in t. That is, $d\theta$ and dt have opposite signs, and so we use the negative sign with the square root and write

$$dt = -\sqrt{\frac{L}{2g}} \cdot \frac{d\theta}{\sqrt{\cos(\theta) - \cos(\alpha)}}.$$

For the complete descent, we have t going from 0 to T as θ goes from α to 0, and so, integrating,

$$\int_0^T dt = -\sqrt{\frac{L}{2g}} \int_\alpha^0 \frac{d\theta}{\sqrt{\cos(\theta) - \cos(\alpha)}},$$

or

$$T = \sqrt{\frac{L}{2g}} \int_0^\alpha \frac{d\theta}{\sqrt{\cos(\theta) - \cos(\alpha)}}.$$

From the trigonometric half-angle identities,

$$\cos(\theta) = 1 - 2\sin^2\left(\frac{1}{2}\theta\right)$$

$$\cos(\alpha) = 1 - 2\sin^2\left(\frac{1}{2}\alpha\right),$$

and so

$$T = \sqrt{\frac{L}{2g}} \int_0^\alpha \frac{d\theta}{\sqrt{2\sin^2\left(\frac{1}{2}\alpha\right) - 2\sin^2\left(\frac{1}{2}\theta\right)}}$$

$$= \frac{1}{2}\sqrt{\frac{L}{g}} \int_0^\alpha \frac{d\theta}{\sqrt{\sin^2\left(\frac{1}{2}\alpha\right) - \sin^2\left(\frac{1}{2}\theta\right)}}.$$

Since α is a constant, then so is $\sin\left(\frac{1}{2}\alpha\right)$, which I'll now write as simply k. Thus,

$$T = \frac{1}{2}\sqrt{\frac{L}{g}}\int_0^\alpha \frac{d\theta}{\sqrt{k^2 - \sin^2\left(\frac{1}{2}\theta\right)}}, \qquad k = \sin\left(\frac{1}{2}\alpha\right).$$

Next, if we make the change of variable from θ to β, defining β to be such that

$$\sin(\beta) = \frac{\sin\left(\frac{1}{2}\theta\right)}{\sin\left(\frac{1}{2}\alpha\right)} = \frac{\sin\left(\frac{1}{2}\theta\right)}{k},$$

then as θ varies from α to 0, we have $\sin(\beta)$ varying from 1 to 0, i.e., β varies from 90° ($= \pi/2$ radians) to 0. Differentiating the relationship $\sin\left(\frac{1}{2}\theta\right) = k\,\sin(\beta)$ with respect to θ, using the chain rule, we have

$$\frac{1}{2}\cos\left(\frac{1}{2}\theta\right) = k\cos(\beta)\frac{d\beta}{d\theta},$$

or, solving for $d\theta$,

$$d\theta = \frac{2k\cos(\beta)}{\cos\left(\frac{1}{2}\theta\right)}d\beta.$$

Next, we use $\sin\left(\frac{1}{2}\theta\right) = k\,\sin(\beta)$ to write (since $\sin^2 + \cos^2 = 1$)

$$\cos\left(\frac{1}{2}\theta\right) = \sqrt{1 - k^2\sin^2(\beta)}.$$

Also, since

$$\cos(\beta) = \sqrt{1 - \sin^2(\beta)} = \sqrt{1 - \frac{\sin^2\left(\frac{1}{2}\theta\right)}{k^2}} = \frac{1}{k}\sqrt{k^2 - \sin^2\left(\frac{1}{2}\theta\right)},$$

then

$$d\theta = \frac{2k \cdot \frac{1}{k}\sqrt{k^2 - \sin^2\left(\frac{1}{2}\theta\right)}}{\sqrt{1 - k^2 \sin^2(\beta)}} \quad d\beta = 2\frac{\sqrt{k^2 - \sin^2\left(\frac{1}{2}\theta\right)}}{\sqrt{1 - k^2 \sin^2(\beta)}} \quad d\beta.$$

Inserting this into the integral for T (and modifying the limits to fit the new variable of integration, β) we arrive at our final answer:

$$T = \sqrt{\frac{L}{g}} \int_0^{\frac{\pi}{2}} \frac{d\beta}{\sqrt{1 - k^2 \sin^2(\beta)}}, \qquad k = \sin\left(\frac{1}{2}\alpha\right).$$

None of the math in this appendix had been invented yet in Galileo's time, and so his approach to studying the descent time along a circular path was by the different, approximate method discussed in section 6.1. And even after the integral for T had been derived (by 1700), nobody knew how to evaluate it. Not even the genius of Euler or Newton could see how to do it. All of the attempts to express the integral in terms of the then-known elementary functions (e.g., logarithms, powers, exponentials, trigonometric functions) failed. It wasn't until more than 150 years after Galileo's death, with the work of the French mathematician Andrien Marie Legendre (1752–1833), that it was appreciated that the failure is due to impossibility; the integral for T represents an entirely *new* function! The integral, called the *complete elliptic integral of the first kind*, has been numerically evaluated for numerous values of α and can be found in many mathematical tables. For example, if $\alpha = 90°$ (the bead descends along a full one-quarter arc of a circle), then

$$T = 1.8541 \sqrt{\frac{L}{g}}.$$

Appendix F.

The Area Enclosed by a Closed Curve

Imagine that the closed, non-self-intersecting curve C shown in figure F1 is described by the parametric equations

$$x = x(t)$$
$$y = y(t),$$

where the parameter t denotes time. That is, at time $t = 0$ we imagine a point particle is at A, the location of the *left* vertical tangent to C. Then, as time increases, the particle moves according to the parametric equations, thereby tracing out the curve C until it returns to A at time $t = T_A$. We also define the time $t = T_B$ as the time at which the moving particle reaches B, the location of the *right* vertical tangent to C. (The two vertical tangent lines are called C's *lines of support*.) To be very specific, let's also assume that the parametric equations describe a *clockwise* motion of the particle. The claim, then, is that the *area enclosed* by C is given by

$$\text{enclosed area} = \frac{1}{2} \int_0^{T_A} (y\dot{x} - x\dot{y}) \, dt,$$

where $\dot{x} = dx/dt$ and $\dot{y} = dy/dt$, where I am using Newton's dot notation to denote a *time* derivative.

Before deriving this result, let me give you an example of its use. For the special case of C, a circle of unit radius centered on the origin (see figure F2), the parametric equations of C are

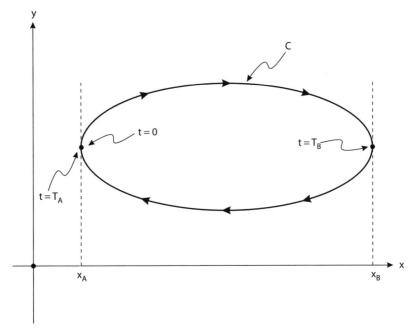

FIGURE F1. A closed, non-self-intersecting curve C is the path of a moving point.

$$x(t) = -\cos(t)$$
$$y(t) = \sin(t),$$

with $T_A = 2\pi$ and $T_B = \pi$. This describes clockwise motion, with $A = (-1, 0)$ and $B = (1, 0)$. Now, since $\dot{x} = \sin(t)$ and $\dot{y} = \cos(t)$, then the claim says

$$\text{enclosed area} = \frac{1}{2} \int_0^{2\pi} \left[\sin^2(t) + \cos^2(t) \right] dt = \frac{1}{2} \int_0^{2\pi} dt = \frac{2\pi}{2} = \pi,$$

which is, indeed, the area of the circle.

In this example, C actually cuts through all four quadrants of the xy-plane, but in the proof I'll assume C lies entirely in the first quadrant (as drawn in figure F1). This assumption is strictly for convenience, however, and when we are done you'll see it will in no way weaken the result. And finally, in figure F1, I have drawn C as a closed *convex* curve and so for each value of x there are

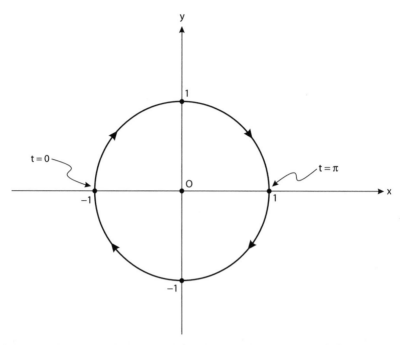

FIGURE F2. A unit circle traced out by a moving point in time interval 2π.

at most two values of y. Our result is easy to extend to *concave* curves, too, however, by dividing such a C up into sections, each of which has just two lines of support. For all curves but for those that mathematicians call pathological (i.e., diseased!) this sectioning can always be done in a finite number of steps. This issue of concavity is actually not important for us in this book, as we will use the result only (in chapter 6) to solve the ancient isoperimetric problem of determining the closed curve of given perimeter that encloses the greatest area. And, as shown in section 2.2, the solution *must be convex*, just as drawn in figure F1.

So, we begin. If we write the integral

$$\int_{x_A}^{x_B} y(x)\,dx,$$

we get the area between the top of C and the x-axis (because I have assumed the x-axis lies completely below C) *if* we use $y(x)$ for the upper half of C. Writing

$$dx = \frac{dx}{dt} \, dt,$$

the integral becomes

$$\text{area under top half of } C = A_1 = \int_0^{T_B} y(t) \, \frac{dx}{dt} \, dt,$$

where the limits on the integral are changed to match the new integration variable t. That is, $x = x_A$ at $t = 0$ and $x = x_B$ at $t = T_B$. Next, if we write the integral

$$\int_{x_B}^{x_A} y(x) \, dx,$$

we get the *negative* of the area (because $x_B > x_A$) between the bottom of C and the x-axis *if* we use $y(x)$ for the lower half of C. Thus,

$$\text{area under bottom half of } C = A_2 = -\int_{T_B}^{T_A} y(t) \, \frac{dx}{dt} \, dt.$$

Now, the area *inside* of C, i.e., the enclosed area, is simply $A_1 - A_2$, and so

$$\text{area enclosed by } C = \int_0^{T_B} y(t) \, \frac{dx}{dt} \, dt + \int_{T_B}^{T_A} y(t) \, \frac{dx}{dt} \, dt$$

$$= \int_0^{T_A} y\dot{x} \, dt.$$

This expression is the answer, as it stands, to the question of what area is enclosed by C. It does have the property, however, of appearing to treat x and y differently (one is differentiated and the other is not), despite the fact that a choice of coordinate system is arbitrary. We can get our answer to look symmetrical (to treat x and y the same) with the following last step.

Imagine we rotate the coordinate axes (and C) counterclockwise by 90°, to arrive at figure F3. That is, y is replaced with x and x is replaced with $-y$. Since the enclosed area is a *physical invariant* unaffected by a particular choice of coordinate axes, our result for the enclosed area must be

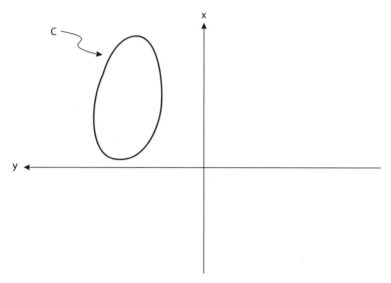

FιGURE F3. Coordinate rotation of C does not affect the enclosed area.

$$\text{area enclosed by } C = \int_0^{T_A} x(-\dot{y}) \, dt = -\int_0^{T_A} x\dot{y} \, dt.$$

(If we had rotated *clockwise* by 90°, then we would have replaced y with $-x$ and x with y, which would lead to the same conclusion. Can you see what happens with a 180° rotation, either clockwise or counterclockwise? Then we get our original expression back, which, while not wrong, is not useful.) Thus, adding this expression to the original expression says

$$\text{\textit{twice} the area enclosed by } C = \int_0^{T_A} y\dot{x} \, dt - \int_0^{T_A} x\dot{y} \, dt,$$

or, at last, the symmetrical result

$$\text{area enclosed by } C = \frac{1}{2} \int_0^{T_A} (y\dot{x} - x\dot{y}) \, dt.$$

This result can easily be shown to be invariant under coordinate axes translation and/or rotation, which of course merely says *area* is (as mentioned already) a physical invariant.

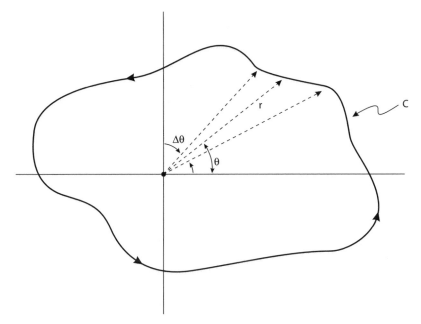

FIGURE F4. The curve C is traced out by the end of a rotating radius vector.

This important result is *so* important that seeing an alternative derivation is not a waste of time. It is, in fact, a result that is often not derived even *once* in first treatments of the calculus of variations, with authors usually writing something like "see any advanced calculus text for a derivation." I don't like that approach, and so let's do it here *again*.

We begin anew with the curve C expressed in polar coordinates this time, i.e., a general point on C is located by drawing the radius vector from the origin to that point, at angle θ with length r, as shown in figure F4. (Without loss of generality, we imagine the origin of our coordinate system is inside, i.e., is surrounded by) C. Then, as θ varies through a total change of 2π radians, the varying length r of the radius vector causes the tip of the radius vector to trace out C in a *counter*clockwise sense.

Since the little "triangle" swept over by the radius vector, through a tiny angular change of $\Delta\theta$, has a base of $r\Delta\theta$ and a height of r, then its area is $\Delta A = \frac{1}{2} r^2 \Delta\theta$. As we let $\Delta\theta \to 0$, we have $\Delta A \to dA$, and so the total area enclosed by C is

$$A = \int dA = \frac{1}{2} \int_0^{2\pi} r^2 \, d\theta.$$

Now, in terms of rectangular coordinates, we have the familiar relations

$$x = r \cos(\theta)$$

$$y = r \sin(\theta),$$

and so the *total* differentials dx and dy are, in terms of the partial derivatives,

$$dx = \frac{\partial x}{\partial r} \, dr + \frac{\partial x}{\partial \theta} \, d\theta = \cos(\theta) \, dr - r \sin(\theta) \, d\theta$$

$$dy = \frac{\partial y}{\partial r} \, dr + \frac{\partial y}{\partial \theta} \, d\theta = \sin(\theta) \, dr + r \cos(\theta) \, d\theta.$$

From these expressions we can now write

$$x(dy) - y(dx) = [r \cos(\theta)] \, [\sin(\theta)dr + r \cos(\theta)d\theta]$$

$$- [r \sin(\theta)] \, [\cos(\theta)dr - r \sin(\theta)d\theta],$$

which, after expansion and the obvious simplifications, reduces to just $r^2 d\theta$.

So, if $\theta = 0$ at time $t = 0$ and if $\theta = 2\pi$ at time $t = T_A$, we have (the symbol $\int_C$ means the integral is completely around the curve C)

$$A = \frac{1}{2} \int_C (x \, dy - y \, dx) = -\frac{1}{2} \int_0^{T_A} \left\{ y \frac{dx}{dt} - x \frac{dy}{dt} \right\} dt$$

$$= -\frac{1}{2} \int_0^{T_A} (y\dot{x} - x\dot{y}) \, dt.$$

This is just what we got in the first derivation, except for the sign. Remember, however, that now we are moving around C in the *counter*clockwise sense, opposite to the sense of travel in the first derivation. So all is, indeed, consistent. Two very different approaches, with the same result, which should add confidence in our minds that we have a correct result.

Appendix G.
Beltrami's Identity

If we multiply through the Euler-Lagrange equation of section 6.4 by y', we get

$$y' \frac{\partial F}{\partial y} - y' \frac{d}{dx}\left(\frac{\partial F}{\partial y'}\right) = 0, \qquad y' = \frac{dy}{dx}$$

where, in general, $F = F\{x, y(x), y'(x)\}$. Now, the change, dF, in F as we allow each of the three explicit variables (x, y, y') to change is given in terms of the partial derivatives of F by

$$dF = \frac{\partial F}{\partial y}\, dy + \frac{\partial F}{\partial y'}\, dy' + \frac{\partial F}{\partial x}\, dx.$$

Or, dividing through by dx, we have

$$\frac{dF}{dx} = \frac{dy}{dx} \cdot \frac{\partial F}{\partial y} + \frac{dy'}{dx} \cdot \frac{\partial F}{\partial y'} + \frac{dx}{dx} \cdot \frac{\partial F}{\partial x}$$

$$= y' \frac{\partial F}{\partial y} + y'' \frac{\partial F}{\partial y'} + \frac{\partial F}{\partial x},$$

assuming y is twice differentiable with respect to x. That is,

$$y' \frac{\partial F}{\partial y} = \frac{dF}{dx} - y'' \frac{\partial F}{\partial y'} - \frac{\partial F}{\partial x}, \qquad y'' = \frac{dy'}{dx} = \frac{d^2 y}{dx^2}.$$

Substituting this expression for $y'(\partial F/\partial y)$ into the first equation above gives

$$\frac{dF}{dx} - y'' \frac{\partial F}{\partial y'} - \frac{\partial F}{\partial x} - y' \frac{d}{dx}\left(\frac{\partial F}{\partial y'}\right) = 0.$$

If we now notice that

$$\frac{d}{dx}\left\{F - y'\frac{\partial F}{\partial y'}\right\} = \frac{dF}{dx} - y''\frac{\partial F}{\partial y'} - y'\frac{d}{dx}\left(\frac{\partial F}{\partial y'}\right),$$

then we see that

$$-\frac{\partial F}{\partial x} + \frac{d}{dx}\left\{F - y'\frac{\partial F}{\partial y'}\right\} = 0.$$

At this point we simply have an alternative form for the Euler-Lagrange equation. But, if we now suppose that F has no explicit dependence on x, i.e., if we suppose that

$$\frac{\partial F}{\partial x} = 0,$$

then we can write

$$\frac{d}{dx}\left\{F - y'\frac{\partial F}{\partial y'}\right\} = 0.$$

This is immediately integrable to give

$$F - y'\frac{\partial F}{\partial y'} = \text{constant.}$$

This is *Beltrami's identity* of 1868, a partially integrated form of the Euler-Lagrange equation for the special (but important) case when F does not depend explicitly on x. This condition is satisfied in a number of historically important problems, and great use of the Beltrami identity is made in chapter 6.

Appendix H.

The Last Word on the

Lost Fisherman Problem

At the end of Chapter 1, I challenged you to find a solution path that is even better (shorter) than the one that gets the fisherman back to shore in no more than 6.9953 miles. Consider the path shown in figure H1, which is but a slight variation of the 6.9953-mile path of figure 1.12; two straight-line segments have been added to the beginning and the end of the circular portion.

Once again, the fisherman rows at some arbitrary angle θ to an *assumed* straight one-mile path to shore. Then, looking back along the path he has just rowed, he turns through an angle of $90° - \theta$ and rows a distance of $\sin(\theta)\sqrt{1 + \tan^2(\theta)}$. This puts him, as shown in figure H1, at an angle of 2θ from the assumed direct path to shore (as well as once again one mile from his starting position). He then rows in a circular path of radius one mile until he is once again at angle 2θ with respect to the assumed direct path to shore. That is, as measured from the *assumed* path to shore, he swings through an angle of $360° - 4\theta$. Finally, he then rows straight ahead along the tangent to the circle at the end of the circular portion of his path. After rowing a *maximum* of $\sin(\theta)\sqrt{1 + \tan^2(\theta)}$, he is sure to arrive at the shore, because the original $(1 + 2\pi)$-mile path lies on or inside this path. The total distance rowed is, with θ measured in radians,

$$L(\theta) = \sqrt{1 + \tan^2(\theta)} + 2\sin(\theta)\sqrt{1 + \tan^2(\theta)} + 2\pi\left(\frac{2\pi - 4\theta}{2\pi}\right)$$

$$= 2\pi - 4\theta + [1 + 2\sin(\theta)]\sqrt{1 + \tan^2(\theta)}.$$

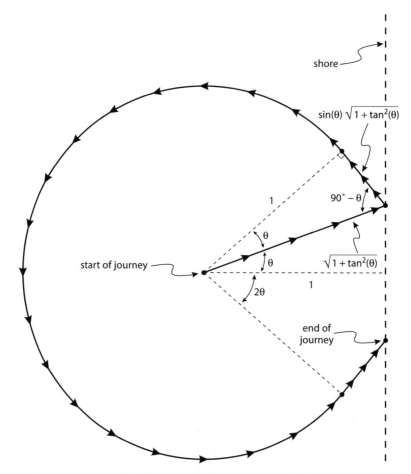

FIGURE H1. The lost fisherman problem, one last time: is *this* the shortest path possible?

We could use a computer to numerically find the θ that minimizes $L(\theta)$ but, somewhat surprisingly, setting $dL/d\theta = 0$ in this more complicated case now gives an analytically solvable equation! If you work through the algebra, you should arrive at the quadratic equation

$$4\sin^2(\theta) + \sin(\theta) - 2 = 0,$$

which has the solution

$$\sin(\theta) = \frac{\sqrt{33} - 1}{8}.$$

Thus, $L(\theta)$ is minimized (do you see why the extrema is *not* a maximum?) when $\theta = 0.63487$ radians ($= 36.375°$), at which value $L = 6.4589$ miles, an impressive 11.3% less than the original ($2\pi + 1$) miles $= 7.2832$ miles calculated in chapter 1.

Appendix I.

Solution to the New Challenge Problem

In figure I1 I've drawn a triangle with side lengths a, b, and c. The three solid lines represent the bisector lines of the vertex angles (dividing the vertex angles into the half-angles α, β, and γ), and these bisector lines meet (as stated in the hint* given in the preface to the paperback edition of the book) at a common point, the point P in the figure. From P I've then dropped perpendiculars (shown as dashed lines) to the three sides. This immediately explains where the three pairs of equal lengths (marked as x, y, and z) come from, which in turn gives us the equations

$$a = z + y,$$

and

$$b = x + z,$$

---

* The proof is elementary. Suppose we call angle A the interior vertex angle formed by sides a and b, angle B the interior vertex angle formed by sides b and c, and angle C the interior vertex angle formed by sides c and a. Then, every point on the bisector line of angle A is equidistant (by symmetry) from a and b, and every point on the bisector line of angle B is equidistant (again, by symmetry) from b and c. Thus, the point P where those two bisector lines cross—and they *must* cross since they are not parallel lines—is a point on *both* bisector lines and so is equidistant from a and b as well as equidistant from b and c; point P is equidistant from a, b, **and** c. That means, in particular, that P is equidistant from a and c *and so* P *is a point on the bisector line of angle* C. That is, *all three bisector lines* of the interior vertex angles intersect at P.

and

$$c = x + y.$$

While it may not be (almost certainly isn't) obvious at this point why we would be interested in doing so, we can now use these three equations to show that

$$a + b - c = (z + y) + (x + z) - (x + y) = 2z,$$

and

$$b + c - a = (x + z) + (x + y) - (z + y) = 2x,$$

and

$$a + c - b = (z + y) + (x + y) - (x + z) = 2y.$$

You'll see, soon, how these expressions will be of great use to us.

Now, a brief pause to establish a result we'll need to finish our analysis. If u and v are any two numbers, then it is clear that

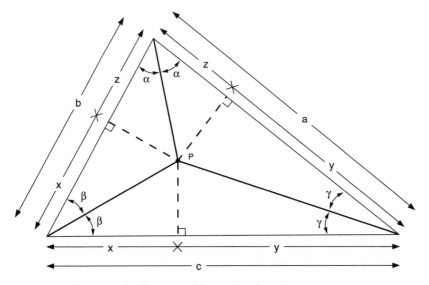

FIGURE I1. The new challenge problem triangle.

$$(u - v)^2 \geq 0$$

or,

$$u^2 - 2uv + v^2 \geq 0$$

or,

$$u^2 + 2uv + v^2 \geq 4uv$$

or,

$$(u + v)^2 \geq 4uv.$$

For our problem u and v are both non-negative numbers—they will denote the lengths of two of the sides of any triangle—and so this last inequality says that

$$u + v \geq 2\sqrt{uv}.$$

This is, in fact, the most elementary possible special case of the AM-GM inequality, which is proven in much more generality in appendix A and which is used in numerous places in this book. Okay, back to our original problem.

Using our first three equations, we have

$$abc = (z + y)(x + z)(x + y).$$

But since our above AM-GM inequality tells us that $z + y \geq 2\sqrt{zy}$, $x + z \geq 2\sqrt{xz}$, and $x + y \geq 2\sqrt{xy}$, we have

$$abc \geq (2\sqrt{zy})(2\sqrt{xz})(2\sqrt{xy}) = (2z)(2x)(2y).$$

And then, finally, our earlier equations for $2z$, $2x$, and $2y$ complete the analysis:

$$abc \geq (a + b - c)(b + c - a)(a + c - b),$$

where a, b, and c are the lengths of the sides of any triangle. Q.E.D.

This result is called *Padoa's inequality*, after the Italian mathematician Alessandro Padoa (1868–1937).

Acknowledgments

A number of people have helped me write this book.
Much of the early work was done while I spent a sabbatical year
in the electrical engineering department at the University of Virginia. The UVA math library in Kerchof Hall, with its rich variety
of holdings, was invaluable to me, and I spent a lot of time there.
I thank the Board of Trustees of the University of New Hampshire
for granting me permission to be absent during the academic year
1999–2000, and the Board of Governors of the University of Virginia for allowing me to be present. Nan Collins, at UNH, typed the
entire book (her fifth for me). Two reviewers of the book offered
particularly helpful advice during revisions: Mary Ann Branch Freeman at The MathWorks in Massachusetts (coincidentally, it was The
MathWorks' *MATLAB* that I used to create the computer generated
plots), and the well-known historian of mathematics Jeremy Gray
at the Open University in England. I benefited from a number of
conversations about high school mathematics with my daughter,
Kimberly Stephens, who teaches mathematics at Noble High School
in North Berwick, Maine. All of the non-computer-generated figures
in the book were skillfully redrawn at Princeton University Press
by Christopher L. Brest; after seeing the professional quality of his
work I was convinced that I should never again be allowed to draw
anything other than a straight line (and even then only if I have a
straight-edge *and* the edge is sufficiently dull that I can't accidentally
wound myself). The book was carefully copy-edited at the Press by
Anne Reifsnyder. My editor, Vickie Kearn, is so perfect that if she
and I did not, together, total 66 years at being the spouses of others,
I would be on my knees asking for her hand. And after saying that,
let me hasten to add that this book would not exist, indeed I would
not exist, without the support and understanding of my loving wife

Patricia Ann. There is no other woman on earth (except for maybe Vickie) who would let me intermingle my scholarly writing with the noise and mayhem (speakers cranked to the max) of *Grand Theft Auto 3* at ten o'clock at night. The cats sometimes complain about the racket, but Pat never has. She *is* the best!

Index